Transitions from Digital Communications to Quantum Communications

To my Mother, with my deep gratitude and affection

Series Editor
Guy Pujolle

Transitions from Digital Communications to Quantum Communications

Concepts and Prospects

Malek Benslama
Hadj Batatia
Abderraouf Messai

iSTE

WILEY

First published 2016 in Great Britain and the United States by ISTE Ltd and John Wiley & Sons, Inc.

ISTE Ltd
27-37 St George's Road
London SW19 4EU
UK

www.iste.co.uk

John Wiley & Sons, Inc.
111 River Street
Hoboken, NJ 07030
USA

www.wiley.com

Library of Congress Control Number: 2016940263

British Library Cataloguing-in-Publication Data
A CIP record for this book is available from the British Library
ISBN 978-1-84821-925-0

Contents

Foreword

Four works dedicated entirely to satellite communications: this is the challenge set by Professor Malek Benslama of the University of Constantine, who understood that a new discipline was in the process of taking shape.

He demonstrated this by organizing the first international symposium on Electromagnetism, Satellites and Cryptography at Jijel in June 2005. The success of congress, surprising for a first-time event, shows that there was a need to gather, in a single place, specialists with skills that are sometimes very removed from one another. The 140 papers accepted concerned systems for electromagnetic systems as well as circuit and antennae engineering and cryptography, which is very often based on pure mathematics. A synergy between these disciplines is necessary to develop the new field of activity that is satellite communication.

The emergence of new disciplines of this type has already taken place before: for electromagnetic compatibility, it was as necessary to know electrical engineering for "driven modes" and "choppers" as electromagnetics ("radiating modes") and to be able to define specific experimental protocols. Further back in time, we saw the emergence of computing which, at the start, lay in the field of electronics and was able, over time, to become independent.

Professor Benslama has the outlook and open-mindedness indispensable for bringing to fruition the synthesis between the skills that coexist in satellite telecommunications. I have known him for 28 years and for me it is a real pleasure to remember all these years of close acquaintance. There has not been a year in which we have not had an opportunity to see one another. For 15 years he worked on the interaction between acoustic waves and semi-conductors. He specialized in resolving piezoelectric equations (Rayleigh waves, creeping waves, etc.), and at the same time was interested in theoretical physics. A doctoral thesis in engineering and then a state thesis crowned his professional achievements. Notably, his examination committee included Jeannine Henaf, then Chief Engineer for the National Center for Telecommunications studies. He was already interested in telecommunications, but also, with the presence of Michel Planat, responsible for research at CNRS, in the difficult problem of synchronizing oscillators.

It is with Planat that he created the path that would lead to quantum cryptography. He made this transformation over 10 years, thus moving without any apparent difficulty from Maxwell's equations to Galois groups. He is now therefore one of the people most likely to dominate all those diverse disciplines that form satellite telecommunications.

I wish, with all my friendly admiration, that these four volumes are with a warm welcome from students and teachers alike.

Henri BAUDRAND
Professor Emeritus
ENSEEIHT

Preface

This book follows on from two other books published by ISTE [BEN 15a, BEN 15b]. I would like to express my respect and gratitude for this publishing house, which makes a decisive contribution to the publishing and distribution of French-speaking scientific authors. I could not forget Professor Guy Pujolle, Director of the *Networks and Telecommunications Series*, in which this book is published. I am infinitely grateful to him.

The idea of transitions from the digital to the quantum field came to me during my stay at Besancon LPMO under the direction of Michel Planat. The discussions that I had with Planat encouraged me to move in the direction of fundamental scientific research. Initially, I met Michel Planat at Toulouse when I defended my state doctorate, which concerned the evolution of surface waves in piezoelectric materials. He suggested that I move towards the non-linear domain in terms of propagation. This then opened the way for me towards solitons in the first instance and then to chaos. Planat was then working on synchronizing surface wave oscillators; he ended by finding an interesting correlation with number theory. This opened the way for formalizing synchronization concepts. Subsequently, his work took a formal, mathematical orientation with the investigation of decoherence phenomena. The quantum aspect was becoming prevalent, and processing tools were drawn from higher math (Galois groups, Lie algebra, non-commutative geometry, etc.). All my respect, admiration and regards go to him.

The act of formalizing concepts is not unique to engineers, as the need for feasibility and practice means that any excessive theorizing has been avoided. Only the evolution of technology requires us to make a theoretical effort to explain, understand and justify this evolution. This book is a contribution to this effort. To this end, we have focused on linking the relevant, practical application to each concept.

Malek BENSLAMA
May 2016

Introduction

Progress recently demonstrated in crystal photonics and nanotechnologies suggests that communications will imminently drift towards quantum communications. An ever-increasing demand for bandwidth and very high speed in digital communications has made research into new systems necessary to be able to meet these requirements. Parallel with purely technological research, and with a view to adaptation, it has become necessary to rethink the very foundations of signal processing [CHO 04, GUP 05, CHA 05].

The first thought that came to mind is the following: what are the practical references available in the literature to justify the benefit of studying this transition, as well as the future developments that have been announced? Three distinct domains will be considered: advances in signal processing, in digital communications and finally those that relate to quantum communications. In fact, it seemed imperative to us to justify the benefit of this study on the basis of practical advances to show the reader that these transitions are not merely a figment of our imaginations but a tangible reality [VAN 05, CHE 07, BAS 08, PIE 14, KAT 06, HUN 10].

To process a signal very thoroughly and completely, some characteristics must be provided, without which any processing can only be approximate [KAN 09]. Classical techniques have become less important in the face of new considerations based on new concepts [AKY 11]. It is mainly signals resulting from RMN, from tomography and the neurosciences [THA 13] that have created this new problem. Thus, it has been necessary to discuss sparse signals [MAR 12, ZHA 13] and neuro-fuzzy signals.

Two signal characteristics should be considered – electric and photonic signals [DEU 04]. In the interests of a homogeneous study it seems useful for us to explore ways to reconcile particle and waves types [FRI 14]. This is a difficult challenge if we dwell on formal concepts; however, experimentally convincing results allow us to hope for a transition [IAF96, ELD01, SZU 13]. It is always tempting to examine these transitions in terms of continuity, which it too restrictive. We will, however, try to take inspiration from the example of the distributions, then the hyper functions that have resolved the continuity handicap and then the differentiability handicap and from the example of Lebesque integration, which has resolved the convergence problem for Riemann or Stieljes integrals [CHU 94, CHU 95a, CHU 95b, SAB 85].

The most sensible way to study the digital to quantum transition consists of relying on published experimental results [DIM 13]. What at first appears to be a figment of the imagination becomes plausible. Two of the most significant examples are superconductivity [OBA 13] and new materials [BRU 14a, BRU 14b], as well as graphene. A third example is provided by Kaspard Klug's theory in virology [SHO 10] which has found an application in entanglements across the quadrangle, the Fano triangle and the icosahedron, which has enabled an important evolution towards the study of Voronoi cells with a direct application for network covers [COQ 01].

Here, an initial evolution consisted of researching new forms of Fourier transform in which information was not only confined to amplitude, but the phase could provide some aspects of information. To do this, the Fourier transform has evolved into hyper-functions such as the Riemann zeta function or Hardy transforms. On the other hand, the Shannon sampling theorem on which the sampling was based was no longer adequate for weak samples. In fact, the Shannon theorem resulted in a full matrix, whereas a random matrix carrying several zeros enabled the initial signal to be found, as long as the matrix had a certain rank. This gave rise to compressed sensing.

The signal/noise ratio, which is found in information theory formulas, as it has been established since Shannon [SHA 48, SHA 49], is a fundamental element for defining the communications' quality.

It has been shown that the signal to noise ratio, so important in information theory, becomes moot for digital communications where symbols modulate the carrier solutions of differential linear equations with polynomial coefficients. On the other hand, new algebraic techniques for estimation enable demodulation according to Fliess [FLI 07].

The possibility that there are links between information theory and quantum mechanics has been examined by various authors [BRI 59, BRU 01, GRE 00].

The principals brought into question are conveyed by works on signal non-stationarity and the generalized Gaussian as shown by [VAR 89]. This has opened the way for signal processing by principal component analysis or Karhunen–Loeve transformation. Gaussian processes have thereafter been processed in the form of Levy processes and sparse Gaussian processes. The links with quantum communications therefore appeared naturally. These links were then accentuated by new developments obtained from geometry via Clifford algebra, which is a natural extension of the analysis and processing of two dimensional signals such as images. In this way, the quaternion signal and the quaternion phase, the signal and the monogenic phase are constructed in geometric algebra or Clifford algebra [TOD 14].

In *Geometric Algebra for Computer Graphics*, John Vince notes that geometric algebra has a natural affinity with Clifford algebra [VIN 08]. This brings us to a natural transition from Clifford algebra to non-Euclidian geometry. He also states the reasons why geometric algebra only emerged in the 21st Century, even though it was discovered in the 19th Century. The works of Clifford and Hestenes were authoritative, leading to Grassmann's work.

David Hestenes' works start with the compliant model that was used in 5D Minkowski space. Applications are numerous in quantum mechanics, in electrodynamics and, in what concerns us, in graphic computing [HES 84]. A remarkable work by the same author tackled spinors [HES 67].

For programmers interested in genetic algorithms (GA), a wide variety of tools is available for systems such as MAPLE and Mathematica, which has resulted in GAIGEN [FON 06]. Practical applications for refining the coprocessor based on Clifford algebra have been created by [FRA 13]. The opening towards higher algebras has thus been demonstrated.

Dorst and Manne give a calculation framework for geometric applications. Their calculation elements are sub-spaces. They specify that the base elements for the calculation are sub-spaces oriented from higher dimensions called "blades" [DOR 01].

The first definition of geometric algebra is introduced – it is an outer product. In this way, the outer product of vectors a_1, a_2, ... a_k is antisymmetric, associative and linear in these arguments. It is written: $a_1 \not\subset a_2 \not\subset \ldots_k$ and called blade k. The algebraic geometry, however, remains a Euclidian geometry.

Our book will begin with a chapter on non-linear signal processing, with a brief reminder of Gumbel's work on probability distributions. We will not expand on the variance and covariance, which have been discussed enough in the literature. On the contrary, we will set out a reflection on stationarity and Bayes inference, which still constitute an element of research. For us, they are an important introductory aspect of tensors in signal processing, which we will extend to quantum signal processing.

The second chapter will tackle non-Gaussian processes. We will give elements of information on principle component analysis, sparse Gaussian processes and generalizations to Levy processes. The links with quantum communications are also given.

The third chapter is devoted to sparse signals and to compressed sensing, a relatively new domain that merits our attention. The density of the results obtained and their applications in LiFi and WiFi networks, leads us to assemble all the information in a specific work that we will propose to our readers next. The fourth chapter hinges on the Fourier transform and is fairly traditional in its formulation. We have judged it useful to include it to reveal the current and future extensions in Clifford algebra and quantum spinors. The fifth chapter relates to the contribution of arithmetic to signal processing and to Gauss Sums in particular. In fact, it seemed interesting to us to refine the notion of the Abelian group in formulating Fourier transforms. The geometric aspect has not been concealed in this text; it will be tackled in the sixth chapter, with an extension towards Voronoi cells.

There are six integrated applications in this text. The first application relates to MIMO systems; it will highlight the notions of distributions, variance, covariance and stationarity. The second application includes elements relating to wavelets, which are much studied in signal processing. The third and fifth applications tackle adaptive radar and SAR radar respectively. Still with radar in mind, we have considered it useful to reveal the belief functions that are the counterpart to probability in signal processing. Finally, a relevant application for geometry is given via the networks.

List of Acronyms

Ad hoc	Wireless network infrastructure
ACK	ACKnowledgement
AOD	Ad hoc On Demand Distance Vector
ARQ	Automatic repeat request
Bcoh	Channel coherence band
BER	Bit Error Ratio
BF	Belief
BGP	Border gateway protocol
BW	Bandwidth
CCA	Clear Channel Detection
CDMA	Code Division Multiple Access
CEF	Clearing Code
CNR	Clutter to Noise Ratio
CMOS	Complementary Metal Oxide Semiconductor
CP	Cyclic prefix
CRC	Cyclic Redundancy Check
CSMA	Carrier Sense Multiple Access
CSMA/CA	CSMA with Collision Avoidance
CSMA/CD	CSMA with Collision Detection
CTS	Clear to Send

DAB	Digital audio broadcasting,
DCF	Distributed Coordination Function
DFP	Direct Form Processing
DFT	Discrete Fourier Transform
DIFS	Distributed Inter Frame Space
DOF	Degree Of Freedom
DS–CDMA	Direct sequence code division multiple access
DSDV	Destination Sequenced Distance Vector
DSSS	Direct Sequence Spread Spectrum
DSP	Digital Signal Processor
DST	Dempster–Shafer theory
DWT	Digital Wavelet Transform
EDCF	Enhanced DCF
EIFS	Extended IFS
FDM	Frequency division multiplexing
FDMA	Frequency Division Multiple Access
FEC	Forward Error Correction
FFT	Fast Fourier transform
FHSS	Frequency Hopping Spread Spectrum
FIFO	First Input First Output
FM	Frequency modulation,
FPGA	Field-programmable gate array,
GAF	Geographic Adaptive Fidelity
GMTI	Ground Moving Target Indicator
GSR	Global State Routing
HDTV	High Definition Television
ICI	Inter-carrier interference
IDFT	Inverse Digital Fourier Transform
IEEE	Institute of Electrical and Electronics Engineers

IF	Improvement factor
IFS	Inter Frame Space
IP	Internet Protocol
JNR	Jammer on noise
JPEG	Joint Photographic Experts Group
LEO	Low Earth Orbit
LLC	Logical Link Control
LOS	Line of Sight
MAC	Medium Access Control
MANET	Mobile Ad hoc Network
Mbps	Megabytes per second
MCM	Multicarrier modulations
MIMO	Multiple Input Multiple output
MMSE	Minimum mean square error
Mod	Modulo
MTI	Moving Target Indicator
MVDR	Minimum Variance Distortionless Response
NMR	Nuclear Magnetic Resonance
OFDM	Orthogonal Frequency Division Multiplexing
OLSR	Optimized Link State Routing
OPNET	Open network technologies
PAMAS	Power Aware Multi Access Protocol with Signaling
PCMCIA	Personal Computer Memory Card International Association
PHY	Physical layer
PRF	Pulse Repetition Frequency
QAM	Quadrature Amplitude Modulation
QCA	Quantum Computer automaton
QCP	Quantum Computer point

QoS	Quality of Service
Qubits	Quantum bit
RAP	Random Access Protocol
RCS	Radar cross section
RNG	Relative Neighborhood Graph
RREP	Route Reply
RREQ	Route Request
RS	Reed Solomon
RSSI	Received Signal Strength Indicator
RTS	Request to Send
SAP	Space adaptive processing
SAR	Syntheticaperture radar
SIFS	Short IFS
SINR	Signal interference noise ratio
SINRLw	Signal interference noise ratio loss
SISO	Simple Input Simple Output
SMI	Sample Matrix Inversion Algorithm
SNR	Signal noise ratio
SPAN	Switched Port Analyzer
STAP	Space time adaptive processing
STAP HR	Space time adaptive processing high resolution
TDMA	Time Division Multiple Access
TFR	Fast Fourier Transform
TOR	Fast Wavelet Transform
TS	Time Slot
UDG	Unit Disk Graph
VLSI	Very large scale integration
Vor	Voronoi
WiFi	Wireless Fidelity

WLAN	Wireless Local Area Network
WMN	Wireless Mesh Network
WSN	Wireless Sensor Network
WRP	Wireless Routing Protocol
ZRP	Zone Routing Protocol
$\otimes$	Cyclic convolution operator
[Bel(A)]	Belief (A)
[Pl (A)]	Plausible (A)
INSAR	Interferometric synthetic aperture radar

PART 1

Theory

1

Non-linear Signal Processing

We will summarize the evolution of signal processing by period, according to the technology then available. The periods range from 1910 until the present day, where technology evolved from vacuum tubes to FPGA, DSP and now QCA. The analytical evolution ranges from the study of transfer functions to inverse linear theory. From 1910 until 1940, Fourier [FOU 27], Laplace [LAP 14], Bode and Nyquist [NYQ 24] were the main authors. From 1940 until 1960, we will cite Gabor [GAB 46], Shannon [SHA 48] and Wiener [WIE 49]. From 1960 until 1980, Oppenheim [OPP 99], Kailath, Kalmann [KAL 60], Slepian [SLE 74] and Cooley and Tukey [COO 65] often appear in published articles. From 1980 until 2000, we will cite Kennedy [KEN 11] and Eberhart [EBE 95].

From 2000 until the present, technologically, the evolution has taken two forms, that of photonic crystals and that of quantum dot cells. [ANG 07] demonstrated the possibility of implanting quantum gates with photonic crystal wave guides. The photons confined in the photonic crystals and the wave guides formed by linear chains doped by default with atoms or quantum dots can generate strong non-linear interactions between the photons, thus enabling one or two qubit of quantum gates to be implemented. The simplicity of the gate-switching mechanism, the experimental feasibility of manufacturing two-dimensional photonic crystal devices and integrating such devices with optoelectronic components offer promising new possibilities for processing quantum and optic information networks.

[HEL 97] studied quantum dot cells (QCP) and demonstrated their convenience for non-linear signal processing. [LEN 97] provided the construction of a computing device with quantum dots. The base block of a QCa is a nanometric cell that contains five quantum dots. Each cell contains two electrons, the two electrons are on opposing dots in the cell due to Coulomb repulsion. The electrons can change place from one dot to another by tunneling, so the electrons can have two different states that correspond to the bits' Boolean values. The quantum point cells have three advantages: a very fast execution speed and infinitely reduced dimensions and energy consumption rates. The theoretical developments are sufficiently advanced. However, manufacturing them poses enormous difficulties. This is due to their quantum aspects, which should be taken into account. The only achievement to date is the quantum dot cellular automaton (QCA) [WHI 07].

Period	*Physics*	*Analytics*	*Names*
1910-1940	Empty tubes, localized circuits	Impulse response, Transfer function, transformation methods	Fourier, Laplace, Bode, Nyquist
1940-1960	Microwave circuits	Statistical concepts (correlation, adapted filters, information theory)	Gabor, Shannon, Wiener
1960-1980	Integrated computer, circuits, optical technology	Digital filters, spectrum estimation, fast Fourier transform, inverse linear theory.	Kailath, Oppenheim, Slepian, Tukey
1980-2000	FPGA, DSP, DSP in particle swarm	Ultra-fast signal processing	Kennedy, Eberhart
2000-2015	QCP, QCA	Quantum signal processing	Helsingius, Whitney, Athas

Table 1.1. *Devices, topics and main discoverers*

The simple QCAs at first reveal many problems in creating actual circuits. Thermal noise can alter the state of different cells, its effect increases with the cell's dimensions and the temperature. Other problems limit the maximal dimensions of individual circuits. Large circuits can themselves commute into undesirable states. However, the recent discovery of adiabatic pipelines seem to compensate this problem [ATH 94]. Large circuits can be divided into many small circuits. The dimension of each small circuit is matched such that the thermal noise does not disrupt the calculation. Another advantage of adiabatic pipelines lies in the manufacturing of delay lines and finite state machines, this being an essential advantage for most algorithms.

The evolution of signal processing to the non-linear domain is motivated by technological advances in CMOS [HUN 10], new perspectives in micro-electronics [LUR 04] and new logic gates [MEY 07] as well as nano-networks [AKY 11]. The main works to tackle signal processing are the following: Kurth describes other approaches to identify non-linearities in time series in geophysics and astrophysics. His work consists of a data pretreatment and tests on signal stationarity and artifacts, for which he provides a robust non-linear method for reducing noise [KUR 94]. [KAT 06] discuss the problems and methodologies for systems and signals, with particular attention on modeling, identifying and processing signals. It reveals some common precepts between technologies. In his work, he summarizes the new directions and predictions on systems and signals. In his published technical note, Kijewski reports the increasing use of frequency conversion time for analyzing and interpreting non-stationary and non-linear signals in a wide range of scientific and technical techniques. He focuses his attention on continuous wavelet transform and mode decomposition in tandem with Hilbert transform. His study evaluates the performance of two approaches in analyzing a large variety of classic non-linear signals. The fundamental difference between the two approaches appears in the instantaneous frequency obtained using Hilbert transform, which characterizes sub- and super-cyclic nonlinearities simultaneously, while the wavelet based on the instantaneous frequency captures the super-cyclic nonlinearities with an additional measure of the bandwidth characterizing the sub-cyclic nonlinearities [KIJ 07].

Among the works applied to the nonlinear analysis of medical signals using data mining, we will cite [BOG 10]. Dougherty's work on determining coefficients to decide on the signal's non-linear character deserves to be mentioned [DOU 00]. Diversity has been handled by Modarres using a Hammerstein filter [MOD 09]. Perez has detailed all the characteristics of Gaussian processes in a review article [PER 13]. As for new developments, the future predictions expected in non-linear signal processing are summarized by Gao [GAO 12].

Finally, we will cite a series of books that tackle non-linear signal processing with a variety of methods [ARC 05, MOO 00, STO 05, PRA 08, ROC 04].

1.1. Distributions

When processing the signal, using distributions is fundamental. However, this usage is not clearly defined: we must work with functions or distributions. Of course this depends on the level of abstraction we are working with. Considering what is known as Dirac comb or pseudo function, the best-adapted formalism is that of distributions. Distribution theory in the mathematical sense of the term has been developed by Schwartz [SCH 51]. The objective has therefore been to generalize the notion of function, in order to give a correct mathematical meaning sense to the objects handled manipulated by physicists, additionally retaining the possibility of carrying out operations such as derivations, convolutions or Fourier or Laplace transforms. This generalization of the notion of function has been pursued in various directions, in particular the notion of the hyper-function, thanks to Sato [SAT 59, SAT 60]. We will examine this notion in detail in the context of the evolution of

Fourier transform, which will enable us to identify a significant aspect of non-linear signal processing [SHA 13]. However, it is Gumbel's works that have enabled the distribution aspect, in the sense of the term relating to probability, to be clarified [GUM 35, GUM 50, GUM 53, GUM 54, GUM 60, GUM 61, GUM 67].

A link exists between the probabilities and the distributions. In fact, if p is a positive function defined over $\Re$, such as:

$$\int_{-\infty}^{+\infty} p(x)dx = 1 \qquad [1.1]$$

knowing the probability density enables the expected value of any function f of the random variable X to be calculated.

$$E(\phi(X)) = \int_{-\infty}^{+\infty} \phi(x)p(x)dx \qquad [1.2]$$

we can say that p defines a regular distribution since p is therefore *a fortiori* summable in $L^1_{loc}(\Re)$, and we have:

$$E(\phi(x)) = \langle p, \phi \rangle \qquad [1.3]$$

Holmes, in his technical report on the role of group theory in signal processing's mathematical foundations, shows the usefulness and benefit of group theory, through the Plancherel formula. The essential idea is to use special unit transformation for compression and decorrelation. The suggestions given show that nonabelian group filters can improve the standard methods of discreet Fourier transform and fast Fourier transform (DFT and FFT) without increasing the calculation's complexity [HOL 87].

1.2. Variance

Variance is a measure that serves to characterize a distribution's dispersion. It indicates how the statistical series or the random variable is dispersed around its mean or its expected value. A zero variance shows that all the values are identical. A small variance is a sign that the values are close to one another, while a high variance is a sign that they are very distant.

1.3. Covariance

Covariance is an extension of variance. Correlation is a normalized form of covariance. This concept is naturally generalized to many variables (random vector) by the covariance matrix.

For the stochastic processes that cover the evolution of a random variable, covariance gives way to the concepts of auto-covariance and autocorrelation, then to the estimation of spectral density for stationary processes.

Space time adaptive processing (STAP) normally requires knowledge of the inverse covariance matrix (ICM) of unwanted signals to detect signals from the desired target. The computational load for generating a reliable inverse covariance matrix prevents an adaptive processing from being implemented in radar systems in real time [DON 05].

1.4. Stationarity

It is often convenient to consider stationarity in a random process, this constitutes a simplification if the process linked to the signal is considered as a stationary process. Suppressing stationary and non-stationary noise in the case of cardiac signals has been developed by [RAH 11].

1.5. Bayes inference

The procedures for statistical inference are applied when the available information is less complete than that usually studied. In this case, the initial information is taken to be a series of probability measures P. With an initial probability measure, an estimation of the corresponding Bayes can be found. The inference procedure recommended, when an initial group of probabilities P is available, is to find a set of estimations corresponding to P. This is called an achievable set of estimates [POT 83]. A practical case is developed in this book on minimal mean square error (MMSE) [LAM 13]. Modern statistical approaches accepted in communication theory use statistical inference in designing and evaluating statistical tests [MET 54]. Farina *et al.* developed an algorithm built around Bayes inference in identifying and tracking radar [FAR 02]. [TAK 05] show that the Jeffreys prior plays an important role in statistical inference.

The statistical inference consists of estimating unknown characteristics of a population from a sample taken from this population. The statistical inference is therefore a set of methods enabling reliable conclusions to be drawn from data from statistical samples.

Methods of statistical inference initially enabled fundamental notions of probability to be deduced, as well as hypothesis tests and confidence intervals as stipulated by [PEA 94], [PEA 01] and [WAL73]. Thereafter, re-sampling techniques came to light with [ULA 04], [EFR 82] and [MIS 57].

Algorithms are sought for processing data or signals using models whose parameters will be adjusted by the statistical use of data. The statistical approach enables robust methods applicable to broad signal categories to be developed with the processing of massive data. On the other hand, modeling complex space–time phenomena is a difficult operation. Many processes can only be partially observed. The latent state's statistical inference (by data assimilation) becomes a problem of paramount importance. Two large categories of space-time models are widespread in the literature [MAN 12]:

– geostatistical models that use a statistical description of covariance functions as a starting point;

– space-time models that enable partial differential equations or their stochastic analogue to be generated from a physical description of the mechanism.

The calculation of the statistical inference can also be formulated as shown by Pereyra *et al.* This has the advantage of resolving the difficulties posed by the Cramer–Rao limit.

The image segmentation obtained by tomography in the case of the brain is based on the differential evolution of the Bayes inference as shown by [WAN 11a]. Another example of the application of Bayes inference is given by [WAN 11b] on the exponential distribution function or risk function in establishing Bayes estimators.

Beyerer gives a generalized Bayesian inference process as shown in Figure 1.1.

Takahashi introduces a universal Bayes which is a Bayesian version of the Martin-Löf test, it establishes a series of theorems linked to Bayes statistical inference in terms of a random sequence [TAK 06].

Another example of inference is given us by mixing models in the statistical segmentation of medical images. A Monte Carlo sampling combined with Markov chains is used [WOO 06].

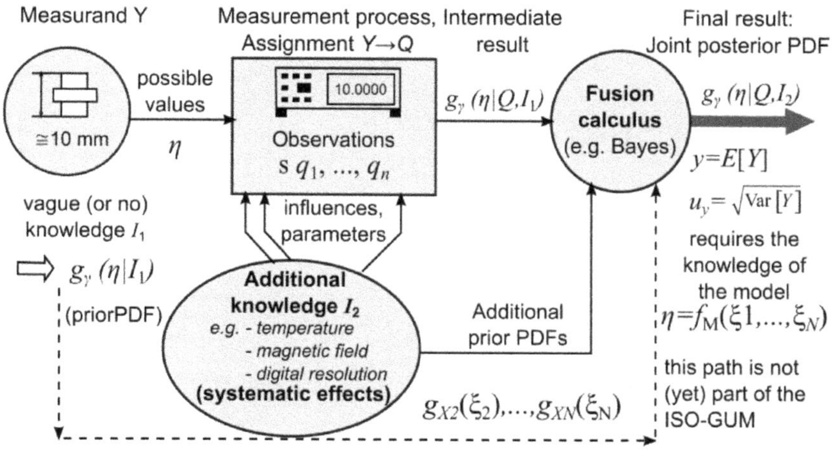

Figure 1.1. *Generalized Bayesian inference process*

Qu and Hu, in a Bayesian framework for generalizing associative functional networks (GAFN), detail a variational Bayes learning algorithm for assimilating later distributions of associative network parameters. This algorithm means statistical inferences can be avoided [QU 07].

An essential application for Bayesian inference is mitigating breakdowns in launching space vehicles. It is cited under the subject "Bayesian anomaly". Contrary to classical inference, the mean estimations differ critically. The difference in the probability of failure in the system's performance is deduced from estimations taken before the test or from real experiments. The probabilities calculated initially differ drastically from the probabilities obtained afterwards [PHI 08].

A second application for Bayesian inference is made on an FPGA architecture. This is made in checking spam in email content. The inference engine uses the logarithmic number system (LNS) to simplify the naive Bayesian calculation. A noise model for the inference engine is developed and the noise limits have been analyzed in order to determine the accuracy of the inference [MAR 08].

Where non-Bayesian signals are concerned, two fairly efficient methods can be used:

– the Cramer–Rao inequality (some call it the Cramer–Rao bound), expresses a lower bound on the variance of an unbiased estimator based on Fisher information.

It states that the reverse of the Fisher information $\Im(\theta)$, of a parameter θ, is a lower bound of the variance of an unbiased estimator of this parameter, written $\hat{\theta}$.

$$\mathrm{var}(\hat{\theta}) \geq \Im(\theta)^{-1} = E\left[\left(\frac{\partial}{\partial\theta}\ln L(X;\theta\right)\right]^{-1} \qquad [1.4]$$

If the model is regular, the Cramer–Rao bound can be written:

$$\Im(\theta)^{-1} = -E\left[\frac{\partial^2}{\partial\theta^2}\ln L(X;\theta\right]^{-1} \qquad [1.5]$$

where $L(X; \theta)$ is the probability function;

– the maximum likelihood method [MOO 99]:

If a sample has produced the finite sequence of numbers x^*_1, x^*_2,x^*_n and if we have chosen to model this situation using a n-sample X_1, …, X_n of random independent variables of the law $\Im(\theta)$, and if choosing the value of the parameter q is the problem we are confronted with, we can consider the event:

$$E^* = \{X_1 = x_1^*, \ldots, X_n = x_n^*\}, \qquad [1.6]$$

Generally:

$$E(x_1,\ldots,x_n) = \{X_1 = x_1,\ldots,X_n = x_n\} = \{X_1 = x_1\} \cap \ldots \cap \{X_n = x_n\} \qquad [1.7]$$

and its probability:

$$L(x_1,\ldots,x_n;\theta) = P_\theta(E(x_1,\ldots,x_n)) = P_\theta(\{X_1 = x_1\} \cap .. | \cap \{X_n = x_n\}) \qquad [1.8]$$

$$P_\theta(\{X_1 = x_1\})..P_\theta(\{X_n = x_n\}) \qquad [1.9]$$

where this last inequality results from the independence hypothesis of the random variables X_i. The idea is that the choice θ^* which it is sensible to make for θ, is that for which this probability is maximal for the values x^*_1,…x^*_n obtained and to ask:

$$\theta^* = Arg\max_\theta\{L(x_1^*,\ldots,x_n^*;\theta)\} \qquad [1.10]$$

If θ exists, it is unique, and it is the value for which $\theta \mapsto L\left(x_1^*,..,x_n^*;\theta\right)$ is maximal. This comes to resolve the following equation in θ:

$$\frac{\partial L}{\partial \theta}\left(x_1^*,..,x_n^*;\theta\right)=0 \qquad [1.11]$$

By definition, the function L_n (x_1,..,x_n ; θ) is:

for $X_i \mapsto L(\theta)$ is called the probability of law L:

$$L_n\left(x_1,...,x_n;\theta\right)=\prod_{i=1}^{n} P_\theta\left(\left\{X_i=x_i\right\}\right) \qquad [1.12]$$

1.6. Tensors in signal processing

Tensor formulations are not usually used in signal processing. This is due to the fact that signal processing practitioners have found solutions on second order tensors using a symbolic matrix notation as an invariant representation. Examples are provided by linear operators, dyadic products (correlation matrixes) and vector functions derived by Jacobian matrixes [RUI 07]. The operation shown most often is Kronecker's tensor product; a high-dimension signal is obtained using two low dimensions. This is commonly used to obtain a separable multidimensional base.

The systematic use of tensorial concepts in signal processing is motivated by the field of higher order statistics. Thus, entities of higher order statistics such as higher order moments and cumulants [GIA 87, CAR 90] and [CAR 91] are higher order tensors. The mathematical framework is based on multilinear algebra, that is to say generalizing matrix algebra to high-order tensors.

Interesting advances have been reported in this context in blind identification and blind source separation. These advances are based on recent works on value decomposition [BAS 07] and linked algebraic approaches, such as principal component analysis and independent component analysis.

In recent years, interest in developing tensorial methods has grown. This interest is boosted by new medical imaging methods, such as the diffusion tensor in magnetic resonance imaging and the need to obtain these tensors and visualize them as images. Currently, detection methods (sensing) provide tensorial data that are usually arranged as sampled, multidimensional signals. However, none of these measures is entirely reliable, since every tensor gives degraded and noisy data.

Processing for multidimensional data generally begins with cutting the tensor into vectors or observation matrixes, so that second order methods are applicable. These methods are essentially based on the covariance matrix, and more recently on higher order statistics. The processed data are then fused to find the initial tensor's dimension [MUT 07].

The tensors produced by diffusion tensor magnetic resonance imaging represent the covariance in a Brownian model. In this physical interpretation, the diffusion tensors should be defined as symmetric and positive. However, this current approach to the statistical analysis of tensor diffusion, which handles these linear items do not take account of the positive symmetry constraint. The difficulty results from the fact that the diffusion tensor space is not the same shape as a vectorial space. It has been demonstrated that diffusion tensor space is a type of curve known as a symmetric Riemann space. Methods have been developed to produce statistics called means and modes variety in this space. It is also shown that these statistics conserve these tensors' geometric properties by including the constraints that are true positive values. Formulating symmetrical space also leads us to a natural definition of diffusion tensors and to new anisotropic measurement methods [FLE 07].

The benefit of measuring anisotropy is also stated in Castano's article [CAS 07]. A useful generalization is also suggested. In fact, his approach is articulated theoretically properties of the space of multivariate normal distributions where it is possible to define a invariant Riemanian and affine metric and to express statistics on the varieties of defined positive and symmetric matrixes. The contribution gives tools for anisotropic filtering and regularizing tensor fields. Real and synthetic diffusion tensor data are validated.

Other very important works on theory have been produced by the following authors [WES 94, KOL 08, LAT 97, LAT 00b, CIC 08, CIC 14a, CIC 14b]. These works define the framework of a new era for tensorial signal processing on the basis of what are called tensor networks. The applications obtained by tensorial signal processing also cover CDMA, radar and wireless communications [NIO 10, LAT 07, ALM 07].

1.7. Processing the quantum signal

Quantum signal processing comprises two parts, classic processing and tensorial processing. For classic processing, we will refer to Eldar's work [ELD 02], in this article, on the basis of existing or new signal processing algorithms.

For tensorial processing, Zanardi *et al.* show that the division of quantum systems into subsystems is dictated by the measures and interactions accessible. The emergence of a multi-part tensorial structure of the state-space and the notion of quantum entanglement are therefore the observables induced. A general algebraic framework is developed, it formalizes the concept of multi-part tensors. Two essential aspects in the quantum domain are analyzed, the quantum information processing and decoherence control [ZAN 04]. Quantum tensor formulation was established by Hardy [HAR 12].

Another aspect that can ensure the transition from the classic to the quantum aspect is tackled by Le Bihan in his works on the contribution to the processing of valued-symbols on non-communicative algebraic structures. These works are articulated on the signals' quaternion character, which ensures the transition from classic to Clifford algebra, which is a fundamental element for developing calculations in quantum mechanics [BIH 11a, BIH 11b, BIH 11c].

The targeted applications are the quantum information processing (quantum memories and calculators).

Processing quantum signals promises to contribute to designing the first generation of quantum information processors, noting what represents the biggest challenge at the moment for pioneers in quantum information: understanding electronic noise [KNI 02, KVE 03].

We are interested in the way in which the signals' quantum properties, arising from such devices, can be interpreted by classic instruments. Experiments will enable us the question of compatibility between quantum processors and the computers of today to be examined more deeply [BEH 94, KEI 06].

2

Non-Gaussian Processes

2.1. Defining Gaussian processes

A process is called Gaussian if all its finite dimensional laws L $(X_{t1}, \ldots X_{tn})$ are Gaussian $(\forall n \in N, \forall t_1, \ldots, t_n \in T)$. In other words, X–$(X_t)_t$ is Gaussian if any linear combination $a_1 X_{t1} + \ldots . + a_n X_{tn}$ follows a Gaussian law (for any n $\in N, t_1, \ldots, t_n \in T; a_1, \ldots, a_n \in R$).

A Gaussian vector's law $(X_{t1}, \ldots, X_{tn})$ is known through its characteristic function by the mean vector $(E[X_{t1}], \ldots, E[X_{tn}])$ and the covariance matrix $(Cov(X_{ti}, X_{tj})_{1 \leq i,j \leq n})$. From this, it is understood that any law of a Gaussian process is obtained as soon as its mean function a(t)–E[X_t] and the covariance operator K(s,t)–Cov(X_s, X_t) are given.

2.2. Non-Gaussian processes

For a non-Gaussian process, the main difficulty lies in characterizing the processes: while in the particular case of Gaussian processes, only the data for the mean functions and the auto-correlation are enough to determine the process completely, this is not the case with non-Gaussian processes. The minimum quantity of information for approaching the non-Gaussian process's actual behavior should include the order one marginal law and the auto-correlation function. However, the order one marginal law is unknown and any simulation problem should be tackled using only the data from a finite number of its moments.

The rate-distortion function R(D) of an information source was introduced by Shannon [SHA 48] to specify the channel capacity required to transmit information from a source with a mean distortion that does not exceed D. The exact rates have been calculated for Gaussian sources using the least squares criteria. For non-Gaussian sources Shannon gives the upper and lower limits on R(D).

The information rates of non-Gaussian processes have been the object of a detailed study by Gerrish and Schultheiss [GER 03]. The authors handle the improvement of information rate estimations for non-Gaussian sources using the least squares criteria. The sources considered are ergodic, and their statistical properties are characterized by a probability density function with n dimensions, a function which is bounded and continuous.

Independent component analysis of the random vector $\overline{x}=(\overline{x}_1,\overline{x}_2, \ldots ,\overline{x}_p)$ comes down to the following linear, instantaneous, noiseless generative model: $\overline{x}=A\overline{s}$, where the components if the vector: $\overline{s}=(s_1,s_2, \ldots ,s_n)$ are mutually independent and the matrix A is of the fixed size pxn. Nevertheless, the following identification conditions should be verified to make it possible to return to the theoretical model. No one source (components or vector) can follow a Gaussian distribution and the rank of matrix A should be equal to the number of sources.

The first condition results from the nullity of moments and cumulants of an order higher than two for a Gaussian distribution. The independence therefore comes down to a simple decorrelation and the statistical independence hypothesis does not allow the Gaussian sources to be separated, however, it is possible to find other non-Gaussian sources.

The second condition requires us to observe at least as much data as there are sources to be identified. This is the case for principal component analysis or Karhunen–Loève transformation. Blind source separation is now a well-known problem. Techniques for processing the antennae networks are based on using second order statistics, but they remain restrictive to. Thus, when initial information on the propagation or the matrix's geometry is not available, the model can be generalized to a blind source separation model. This model assumes the sources are Gaussian and independent [GAL 97].

One application for Gaussian processes has been made in detecting and estimating radar in non-Gaussian environments.

We give some examples of recognizing targets in SAR scenes, detecting moving targets using HR STAP processing (high resolution space time adaptive processing).

2.3. Principal component analysis or Karhunen–Loève transformation

Principal component analysis or Karhunen–Loève transformation is a method from the data analysis or multivariate statistics family that consists of transforming variables linked to one another (correlated) into new variables décor related from one another. These new components are called "principal components" or principal axes.

This technique is used to describe and visualize data and ensure data décor relation. It has an application field in image processing. An application for the technique of extracting principal components in processing satellite images is given by Wang [WAN 09]. Another application for detecting breakdowns in a monitoring process is given by XU [XU 09]. We can also cite an application consisting of noise reduction in Micro-Doppler radar signatures [DU 15], or de-noising in communication signals [KOU 02]. It is possible to partially recover the target signal to separate the target signal from the clutter. This can be carried out using principal component analysis as shown by [AI 15].

Principal component analysis is also useful for reducing estimation error in signal reconstruction problems [KAL 02]. In this case, this analysis is based on using functions specific to the Fisher information operator. In computer vision and image processing, appearance models in subspace are often used for the visual tracking, detection and recognition of objects. For a given class of objects, the subspace appearance model is constructed by identifying a subspace within a space with a broad dimension or pixel intensity. The appearance of new objects can be approached by linear combinations of several or all the basis vectors. These basic vectors are generally formed by principal component analysis [LI 15].

Xie has shown that it is possible to extend this analysis in classifying and extracting sparse principal components [XIE 12]. The combination of analysis in principal components and wavelets enables quantitative signal extraction processes to be analyzed, as suggested by Zhu [ZHU 07]. Limitations and difficulties are however indicated in the processing of biophysical signals [ROT 70].

2.4. Sparse Gaussian processes

We can also envisage sparse Gaussian processes as stated by Bouveyron in his article.

In a non-stationary instance, Karhunen–Loeve representation is used. In linear monitoring applications or nonlinear recursive filtering, the Kalman filter in conjunction with a Gaussian noise model gives the optimal estimation. This case can be extended to a multiple model approach using the multiple interaction models. However, the calculation's complexity and the cost of this approach become inefficient for monitoring or reconstructing surveillance data in radar or maritime monitoring for example. Sequential Monte Carlo methods are more appropriate for monitoring applications that use non-Gaussian measurements [HAD 11].

2.5. Levy process

This is a stochastic process in continuous time, right continuous with left limits, whose increments are stationary and independent.

A stochastic process $x = \{X_t : t \geq 0$ is called a Levy process if:

1) $X_0 = 0$ almost certainly;

2) independent increments: for any $0 \leq t_1 < t_2 < \ldots < t_n < \infty$, $X_{t1} - X_{t2}$, $X_{t3} - X_{t2}$, $X_{tn} - X_{tn-1}$ are independent;

3) stationary increments: for all $s < t$, $X_t - X_s$ is equal in law to X_{t-s};

4) $t \rightarrow X_t$ is almost certainly continuous on the right and limited on the left.

We observe independent and stationary increments and divisibility among its properties.

An interesting application has been created by Pereyra and Batatia [PER 12] in modeling ultrasound echoes in epidermis tissues using aα stable process. In fact, the ultrasound signals diffracted by the tissue are a complex random Levy process. Other works have borrowed the Levy process path, in particular we will cite [SU 10] and [SU 11].

2.6. Links with quantum communications

The links with quantum communications that we wish to introduce are countless. We will content ourselves with some, whose expressions we will detail and we will cite others for information.

The very nature of the wave function, or more precisely its shape, shows a Gaussian process, and a Fourier transform can be added. On the other hand, noting that two wave packets are never orthogonal shows that the wave packets are always correlated.

We will examine this link through the wave function, which is Gaussian, and which matches perfectly to the characteristics of Gaussian processes.

2.6.1. *Wave function in P*

These wave functions are Gaussian wave packets and can be represented by the following equations:

$$\psi(q)=\frac{e^{-\frac{(Q)^2}{4\Delta Q^2}}}{(2\pi)^{\frac{1}{4}}\sqrt{\Delta Q}}e^{-\frac{q^2}{(2\sqrt{2}\Delta Q)^2}+\frac{1}{2}\left(\frac{Q}{\Delta Q}+i\frac{P}{\Delta P}\right)\frac{q}{\Delta Q}} \quad [2.1]$$

Their Fourier transform will give ψ*(p)*:

$$\psi(p)=\frac{1}{\left(2\pi\Delta P^2\right)^{\frac{1}{4}}}e^{-\frac{1}{4}\left(\frac{p-P}{\Delta P}\right)^2+i\frac{Q}{\Delta Q}\left(\frac{p-P}{\Delta P}\right)} \quad [2.2]$$

2.6.2. *The scalar product of two wave packets*

Suppose that there are two Gaussian wave packets, (Q) and ∏(Q), their scalar product is a Gaussian integral and is never null:

$$\langle\varphi|\psi\rangle=\int\varphi^*(q)\psi(q)dq \quad [2.3]$$

Two Gaussian wave packets are never orthogonal. In the case where they have the same standard deviation ΔQ, the previous equation becomes:

$$\langle\varphi|\psi\rangle=\frac{e^{-\frac{(Q)^2\varphi+(Q)^2\psi}{4\Delta Q^2}}}{\sqrt{2\pi\Delta Q^2}}\int e^{-\frac{q^2}{2\Delta Q^2}+\frac{1}{2}\left(\frac{(Q)\varphi+(Q)\psi}{\Delta Q}+i\frac{(P)\varphi+(P)\psi}{\Delta Q}P\right)\frac{q}{\Delta Q}}dq \quad [2.4]$$

$$\langle\varphi|\psi\rangle=\frac{e^{-\frac{(Q)^2\varphi+(Q)^2\psi}{4\Delta Q^2}}}{\sqrt{2\pi\Delta Q^2}}e^{\Delta Q^2\frac{1}{4}\left(\frac{(Q)\varphi+(Q)\psi}{\Delta Q}+i\frac{(P)\varphi+(P)\psi}{\Delta P}\right)\frac{1}{\Delta Q^2}\sqrt{2\pi\Delta Q^2}}$$

$$=e^{-\frac{1}{4}\left(\frac{(Q)\varphi+(Q)\psi}{\Delta Q}\right)^2-\frac{1}{4}\left(\frac{(P)\varphi+(P)\psi}{\Delta P}\right)^2+\frac{i}{2}\frac{(Q)\varphi+(Q)\psi}{\Delta Q}\frac{(P)\varphi+(P)\psi}{\Delta P}}$$

This scalar product is therefore, up to a phase, a Gaussian function of the "distance" of two wave packets in phase space at a scale defined by ΔQ and ΔP. It can therefore be seen that, even if they are never entirely orthogonal, $|\psi\rangle$ and $|\phi\rangle$ are almost orthogonal when $\langle Q\rangle_\psi$ and Q_ϕ differ by some ΔQ (or when $\langle P\rangle_\psi$ and $\langle P\rangle_\psi$ differ by some ΔP).

2.6.3. *Linear superposition of coherent states*

Linear superposition of two coherent states is often known by the name *Schrödinger's cat.* These states in fact illustrate the famous paradox of *Schrödinger's cat.* The paradox, in its initial description, is based on an experimental scenario, placing a cat in a superposition of dead and living states. If everyone knows that it is a dead cat or a living cat, no-one really know that it is a $\frac{1}{\sqrt{2}}$ (dead + alive) cat, a state permitted by in quantum mechanics.

It is found again, in the superposition of two classic well-defined states. For the sake of simplicity, we will limit ourselves to the superposition of states $|\alpha\rangle$ and $\alpha \in R$, $|-\alpha\rangle$ with the other cases appearing there via movement and rotation in the phase space.

The state $|\psi\rangle$ is defined by the superposition:

$$|\psi > \frac{|\alpha>+|-\alpha>}{\sqrt{2+<\alpha|-\alpha>+<-\alpha|\alpha>}} = \frac{1}{\sqrt{2\left(1+e^{-4\alpha^2}\right)}}(|\alpha > +| - \alpha >) \quad [2.5]$$

and can only qualify as *Schrödinger's cat* when states $|\alpha\rangle$ and $|-\alpha\rangle$ are really different, that is to say when $|\langle\alpha|-\alpha\rangle|^2 << 1$. According to [6.5], this approximation is already valid for $|\alpha\rangle 2|$, because of the rapid decrease in the Gaussian function. In this case, the previous equality is rewritten:

$$|\psi >\cong \frac{1}{\sqrt{2}}(|\alpha > +| - \alpha >)$$

This superposition's wave function in Q is the weighted sum of the wave function of each of the terms:

$$\psi(q) = \frac{1}{\sqrt{2(1+e^{-4\alpha^2})}}(< q|\alpha > +< q| - \alpha >) \quad [2.6]$$

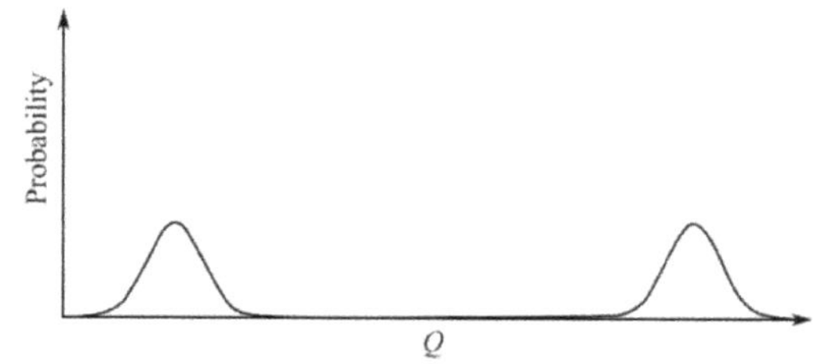

Figure 2.1. *Probability distribution in Q of the linear superposition of two coherent states*

In a case where $\alpha \in R$, this equation becomes:

$$\psi(q) = \frac{1}{(2\pi N_0)^{\frac{1}{4}}\sqrt{2\left(1+e^{-4\alpha^2}\right)}}\left(exp\left[-\left(\frac{q}{\sqrt{4N_0}}-\alpha\right)^2\right]+exp\left[-\left(\frac{q}{\sqrt{4N_0}}+\alpha\right)^2\right]\right) \quad [2.7]$$

where:

N_0: number of electrons;

q: the charge of electrons namely $q = -1.6\ 10^{-19}$ C.

The probability distribution in q associated $|\psi(Q)|^2$ almost corresponds to the two states' mean probability distribution if the Gaussians slightly overlap:

$$|\psi(q)|^2 = \frac{1}{2\left(1+e^{-4\alpha^2}\right)}\left(|<q|\alpha>|^2 + |<q|-\alpha>|^2 + 2Real(<q|\alpha><q|-\alpha>)\right)$$

$$\cong \frac{1}{2}(|<q|\alpha>|^2 + |<q|-\alpha>|^2) \quad [2.8]$$

The probability distribution in the superposed quantum state position is simply the mean of each state's probability distributions. It is not therefore distinguished from a statistical mix where in one instance in two we have state $|\alpha\rangle$ and in one instance in two, state $|-\alpha\rangle$. The difference appears when we look at what happens in P. In fact, the two coherent states' wave functions in P only differ by the phase factor and their probability distributions in P are identical.

$$<p|\pm\alpha> = \frac{1}{\sqrt[4]{(2\pi N_0)}} exp\frac{-p^2}{4N_0} \pm \frac{\alpha p}{\sqrt{N_0}} \quad [2.9]$$

As the probability distributions in P are identical, a statistical mix does not change this distribution, which remains Gaussian. On the contrary, the phase terms will interfere in the quantum superposition and the superposition's wave function in P can be written:

$$\psi(p) = \frac{[2(1+e^{-4\alpha^2})]^{\frac{-1}{2}}}{\sqrt[4]{2\pi N_0}} 2cos\frac{\alpha p}{\sqrt{4N_0}} e^{\frac{-p^2}{4N_0}} \quad [2.10]$$

The probability distribution in P is therefore a Gaussian modulated by a cosine, very different from the Gaussian of the simple statistical mix.

$$|\psi(p)|^2 = \frac{2}{\sqrt{2\pi N_0}\,(1+e^{-4\alpha^2})} e^{-\frac{p^2}{2N_0}} \cos^2\frac{\alpha p}{\sqrt{N_0}}. \quad [2.11]$$

On the contrary, this difference will fade in the presence of noise significant enough to blur fringes. Quantum superposition is fragile.

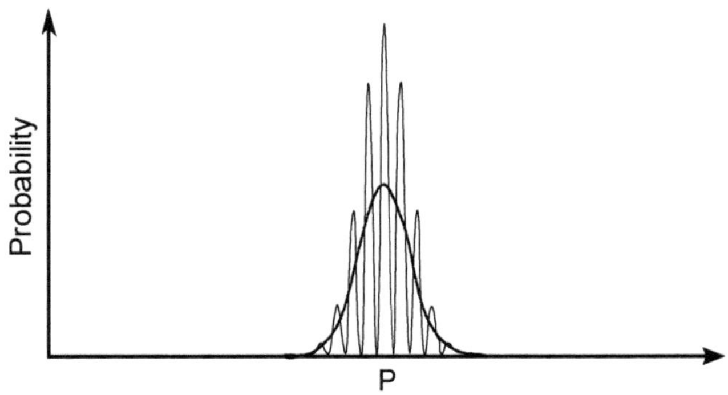

Figure 2.2. *Probability distribution in P compared to a linear superposition (plain line) and a statistical mix (dotted line) of two coherent states*

3

Sparse Signals and Compressed Sensing

3.1. Sparse signals

We owe the most rigorous definition of sparse representation to Mallat. The author thus states that signal processing is faster and simpler in a sparse representation where some coefficients reveal the information we seek. These representations can be constructed by breaking down the signals on the selected elementary wave forms in what is called a dictionary. The ideal sparse transform best adapted to all signals however remains to be found. The discovery of wavelets with orthogonal bases has opened the way for new transformations.

[MAR 12] have given us a very detailed table on sparsity topics. Techniques for understanding the influence of sparsity are described succinctly. The key application domains are sampling, coding, spectral estimation, network processing and estimating multi-path channels.

In the sampling category, the topics addressed range from:

– uniform sampling with a frequency parsimony domain, a type of multi-band parsimony, a time or space information domain, a minimum number of 2xBW-1 samples required, a reconstruction method using a band pass filter and analogue and digital applications;

– non-uniform sampling with a frequency parsimony domain, a type of band pass parsimony, a time or space information domain, a minimum number of samples required from 2xBW-1 in some cases and often BW, a reconstruction method by iterative spline interpolation or by filter bank, and applications in seismology;

– sampling of multi-band signals with a frequency parsimony domain, a type of sparsity obtained by uniting disjoint intervals, a time or space information domain, a minimum number of samples required from 2xBW-1, an iterative reconstruction method, interpolation, and applications in data compression and in radar;

– random sampling, with a frequency parsimony domain, a type of random sparsity, a time or space information domain, a minimum number of samples required from 2xΣ coeff, iterative reconstruction methods with threshold adaptation, applications in lost sample reconstruction and data compression;

– compressed sensing, with a parsimony domain obtained via arbitrary orthonormal transformation, a type of random sparsity, a minimum number of samples required from c.k.log(n/k), a basic or adapted pursuit reconstruction method and applications in data compression.

3.2. Compressed sensing

The traditional approach to reconstructing signals or images of measured data follows Shannon's sampling theorem, which stipulates that the sampling rate should be double the highest frequency. Similarly, the fundamental theorem of linear algebra suggests that the number of samples measured, of a discreet of finite dimension signal should be at least as big as the dimensional length for the signal to be reconstructed. The new theory of compressed sensing, also known as compressed sampling or sparse reconstruction, provides a new fundamental approach for data acquisition, which overcomes the barrier of the Shannon criteria.

Compressed sensing links the empirical observation that many types of signals or images can be approximated well by a parsimonious development in terms of suitable bases by only a small number of non–null coefficients. The compression is obtained by storing only the largest base coefficients. When reconstructing the signal, the coefficients not stored are set to zero.

Consider a one-dimensional discreet-time signal of finite length and real value, this can be seen as an Nx1 column vector in R^N with the elements $x[n]$, $n = 1,2,\ldots,N$. We process an image or a high-dimension data in vector form according to a one-dimensional vector.

Any signal in R^N can be represented in terms of a base of Nx1 vectors $\{\psi_i\}_{i=1}^{N}$. To simplify, suppose that the base is orthonormal. In forming the matrix with the

$N \times N$ base $\Psi = \left[\psi_1 \middle| \psi_2 \middle| .... \middle| \psi_N\right]$ storing the vectors $\{\psi_i\}$ as a column, we can express x as:

$$x = \sum_{i=1}^{N} s_i \psi_i \text{ or } x = \Psi s . \qquad [3.1]$$

where s is the column vector Nx1 of the weight coefficients $s_i = \langle x, \psi_i \rangle = \psi_i^T$ and where T represents the Hermitian transpose. We can say that x and s are equivalent representations of the same signal, with x in the time domain and s in the ψ domain.

We focus on signals that have a sparse representation, where x is specifically a linear combination of *K*-based vectors, with *K* << *N*. We have *K* for those which are non-null, and (*N*-*K*) are null. The sparsity is motivated by the fact that several signals are compressed in the sense that there is a base ψ where the representation has some coefficients that are large and several coefficients that are small.

We give two figures on a sparse discreet-time signal with its discreet Fourier transform as well as the manifestation of parsimony in the frequency domain [MAR 12].

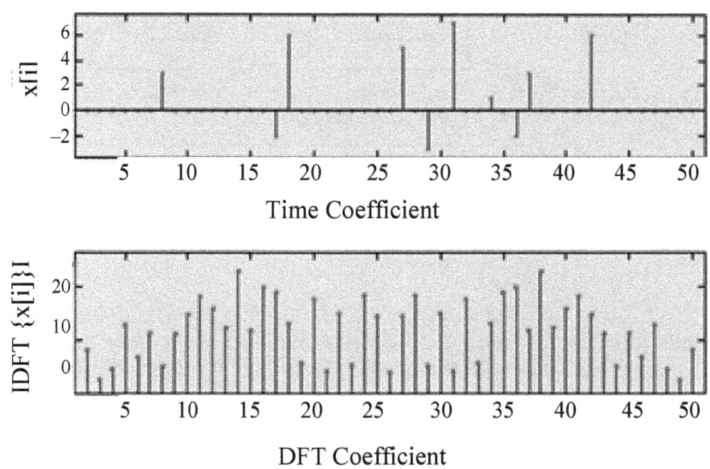

Figure 3.1. *Sparse discreet-time signal with its DFT*

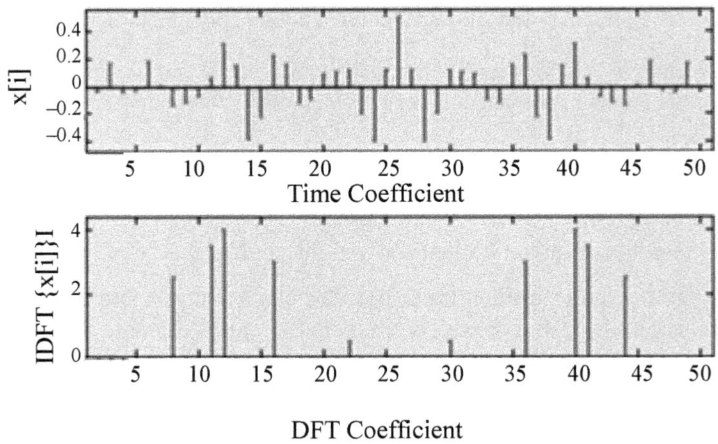

Figure 3.2. *Manifestation of sparsity in the frequency domain*

Figure 3.3 shows an iterative reconstruction method. The mask is an appropriate filter with coefficients of 1 and 0 according to the type of sparsity in the original signal.

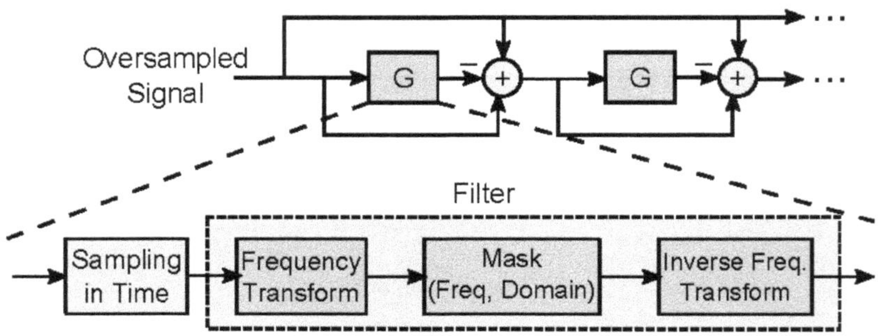

Figure 3.3. *Block diagram of an iterative reconstruction method. The masking is an appropriate filter with coefficients of 1 and 0 according to the type of sparsity in the original signal*

The simulated processing results give us the shape of the signal after processing and the reconstructed signal.

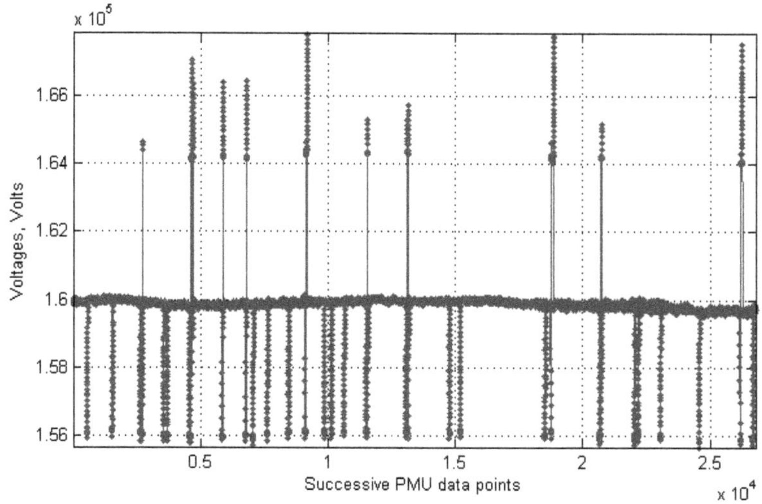

Figure 3.4. *Original Gaussian signal with 100 spikes*

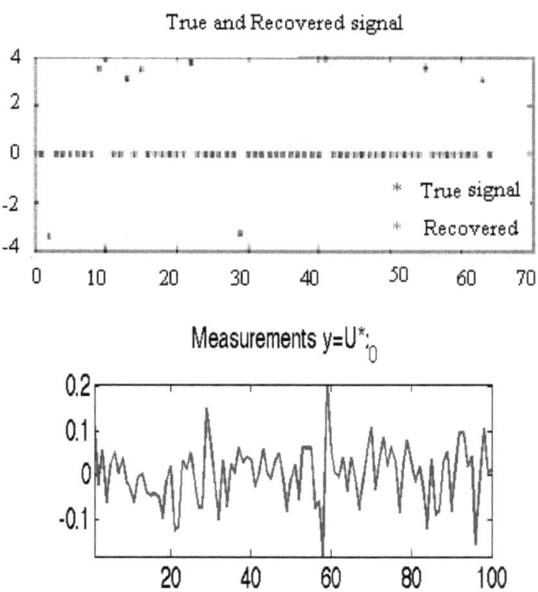

Figure 3.5. *Reconstructed signal and the tally with the coefficients by coefficient of x_0 depending on its reconstruction*

According to [FOR 11], compressed sensing is a new sampling theory which predicts that sparse signals and images can be reconstructed using already known incomplete information. The main feature is that efficient algorithms such as minimization by l_1 can be used for reconstruction.

The following diagram gives an algorithmic overview.

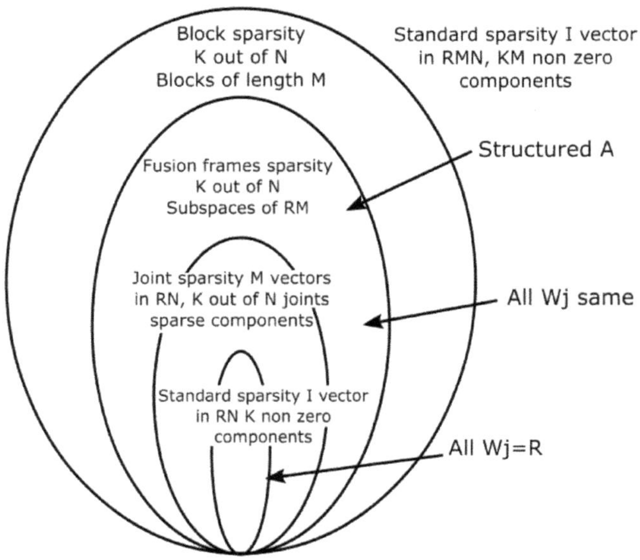

Figure 3.6. *An algorithmic overview of compressed sensing*

3.3. Compressed sensing and quantum signal

In his article on the uncertainty relations for analog signals for a shift invariant, Eldar [ELD 09], by adapting the notion of coherence, defined for finite dictionaries infinite representations of the invariant shift. This class of signals is rich enough to include several interesting special cases such as multiband signals and splines. It shows how to find a sparse decomposition on an over-complete dictionary by resolving a convex optimization problem.

Another work that has drawn much of our attention, as it amounts toan extension of tensors to the quantum domain and states an essential point in breaking dimensionality, is by [VER 14]. This work is based on works previously published by de Lathauer.

4

The Fourier Transform

4.1. The Classic Fourier Transform

The Fourier transform is an operation that transforms an integrable function over **R** into another function describing the latter's frequency spectrum. If f is an integrable function on *R*, its Fourier transform is the function $F(f) = \hat{f}$ given by the formula:

$$F(f): \xi \to \hat{f}(\xi) = \int_{-\infty}^{+\infty} f(x) e^{-i\xi x} dx \quad [4.1]$$

Another formula is often used:

$$F(f): v \to \hat{f}(v) = \int_{-\infty}^{+\infty} f(t) e^{-i2\pi vt} dt \quad [4.2]$$

with t in seconds and the frequency v in Hz.

When processing the signal, we will write t in place of x and ω or 2πv in place of ζ, and these will respectively be the time variables and pulsation variables or frequency variables. We will therefore say that f is in the time domain and that $\hat{f}$ is in the frequency domain.

4.2. The Discreet Fourier Transform and the Fast Fourier Transform

The Fast Fourier Transform (FFT) is a calculation algorithm of the Discreet Fourier Transform (TFD). This algorithm is currently used in digital signal processing to transform discreet data from the time domain into the frequency domain, particularly in spectrum analyzers. Its effectiveness enables filtering to be carried out by modifying the spectrum and using inverse transformation (finite impulse response filter). It is also the basis of the fast multiplication algorithms and digital compression techniques which resulted in the JPEG image format.

It is an algorithm often used to calculate the discreet Fourier transform. It is based on a "divide to rule" approach by means of a recursion. This subdivides a discreet Fourier transform of composite size $n = n_1n_2$ into several discreet smaller size Fourier transforms, n_1 and n_2. This algorithm requires O(n) multiplications by roots of unity, more commonly called rotation factors.

The most classic use of the Cooley–Tukey algorithm is a division of the transform into two parts of identical size $n/2$, with this happening at each stage [COO 65]. This constraint limits the sizes possible, because these should be powers of two. However, factorization remains possible (a principle already known to Gauss). In general, coders try to avoid recursion for questions of performance. It is also possible to mix several types of algorithm during subdivisions.

4.3. The Fourier Transform and hyper-functions

In mathematics as in physics, to tackle a problem in an affine space, we must calculate in a dual space. One method for this, employed in analysis, is the Fourier transform. But this transformation is not local, and lends itself very badly to transition to varieties. Sato's method is on the contrary much better adapted to transition: a real variety becomes more complex and instead of watching the Fourier transform's behavior at infinity, we can watch where the boundary values come from. In technical terms, we watch the contangent bundle (more exactly, manytimesthe cotangent bundle) as the real conormal bundle in the complex. Sato thus defines the hyper-functions'analytic wavefront and so the distributions in particular) [SAT 59, SAT 60]. This is the beginning of micro-local analysis, which has revolutionized analysis [SHA 13]. This analysis is a group of techniques involving Fourier transforms pseudo-differential operators and distributions. The term micro-local implies localization in relation to the position in space and the contangent space directions at a given point [LAU 85]. The notion of the hyper-function is a generalization of distribution theory.

Consider the Dirac comb $\omega = \sum_{n=-\infty}^{+\infty} \delta_{(n)}$ where $\delta_{(n)}$ is the Dirac distribution representing the mass +1 at point n. This is a tempered distribution, with compact support. It is canonically associated with the "Dirac comb hyper-function" given by:

$$w = \frac{1}{2\pi i} \sum_{n=-\infty}^{+\infty} \left[\frac{1}{n-z} \right] \qquad [4.3]$$

We also have a good example of the connection between the Fourier transform in the distribution sense and Heisenberg's uncertainty principle. We will demonstrate it in the following case:

– Suppose $\gamma \in \Re, \gamma > 0$. The Gaussian $t \to e^{-\gamma t^2}$ is very concentrated when $\gamma >> 1$ and very spread when $\gamma << 1$. To be convinced, we begin with function e^{-s^2}, and consider the s set such that $e^{-s^2} \geq \frac{9}{10}$, that is to say that the s $s \in [-c, c]$ with $c^2 = \log(\frac{10}{9})$. Then we change the time scale by posing $s = \sqrt{\gamma}$ t; the s set such that $e^{-\gamma t^2} \geq \frac{9}{10}$ is given by $\sqrt{\gamma} t \in [-c, c]$, which is $t \in \left[-\frac{c}{\sqrt{\gamma}}, \frac{c}{\sqrt{\gamma}} \right]$, an interval even smaller than γ is large, and even larger than γ is small.

– The Fourier transform of a Gaussian is also a Gaussian.

4.4. Hilbert Transform

4.4.1. *Definition*

The Hilbert transform, written H, of a real variable is a linear transformation that enables a real signal to extend into the complex domain so that it verifies the Cauchy–Riemann equations.

$$\hat{s}(t) = H\{t \quad\}(t) = (h * s)(t) = vp\left\{ \int_{-\infty}^{+\infty} s(\tau) h(t-\tau)\, d\tau \right\} = \frac{1}{\pi} vp\left\{ \int_{-\infty}^{+\infty} \frac{s(\tau)}{t-\tau} d\tau \right\} \qquad [4.4]$$

where:

$$h(t) = \frac{1}{\pi t}$$

and: $$vp\left\{\int_{-\infty}^{+\infty} s(\tau)h(t-\tau)\,d\tau\right\} = \lim_{\varepsilon\to 0}\left\{\int_{-\infty}^{t-\varepsilon} s(\tau)h(t-\tau)d\tau + \int_{t+\varepsilon}^{+\infty} s(\tau)h(t-\tau)\,d\tau\right\} \qquad [4.5]$$

Vp is the abbreviation of the Cauchy principal value.

A new, fast algorithm for parallel networks enables the discreet Hilbert transform to be calculated. This algorithm offers an appropriate expression for developing a systolic VLSI network for the Hilbert transform. The architectures suggested have the characteristics of massive parallelism, a regular data stream, a modular nature and local interconnections. These networks offer a high calculation speed for the Hilbert transform for applications in real time signal processing [PAD 97]. A generalization of the Hilbert transform in fractal form as well as its physical interpretation is suggested [PEI 98]. Detection of the edges of digital images has also been checked [PEI 00].

The analytic part of a signal $f(t)$ is obtained by suppressing the content of the negative frequency f, that is to say by suppressing the negative part of Fourier transform $\tilde{f}$. In the time domain, the construction of the analytic part is based on the Hilbert transform, f of $f(t)$. A generalization of the definition of the Hilbert transform to obtain the analytic part of a signal, which is associated with its fractional Fourier transform, is thus obtained. It is also shown that the generalized Hilbert transform has similar properties to ordinary Hilbert transform, but it lacks the fractional Fourier transform's semigroup property [ZAY 98].

It has been shown that the Hilbert transform is stable in relation to the energy norm, but its causal form is not [BOC 99].

Inverting the finite Hilbert transform is an essential step in some image processing and tomography algorithms. A new method has been suggested to make the Hilbert transform inverse, which only requires data on a finite interval. This method is more precise and less sensitive to noise than the collocation method [SID 05]. In some technical processes, the signals are often modulated in a complex manner; this is the case with an ultrasound wave passing through a fluid on which stochastically distributed turbulences are superposed. Signal demodulation can be carried out by digital subsampling to separate narrow sidebands from the carrier frequency.

The signal can be separated in reality and imagination by the Hilbert transform [LIN 06].

The Hilbert transform has also been applied in the geoscience domain. This is achieved using Clifford analysis applied to a sphere. To obtain a spectral characterization, the transform has been broken down into in spherical harmonics [RA 06].

4.5. Clifford algebra and the Fourier Transform

Binz and coauthors have unified the structure of signal theory based on holography and quantum information theory [BIN 03]. Heisenberg groups and their known Lie algebras, like H algebras are concepts that can be implemented technically. Heisenberg algebras are fundamental structures for modeling information processing phenomena.

In an article published after his death, Clifford defines the algebra that carries his name as a generalization of exterior Grassmann algebra and Hamilton's quaternions [CLI 78]. His aim was to create a structure including algebraic and geometric properties in Euclidian space, which he called algebraic geometry. This algebra's importance lies in the fact that it incorporates in a single structure both the interior product and the exterior product, called a Grassmann vector product. Classic examples of Clifford algebra are real numbers, complex numbers and quaternions. The γ matrices introduced by Dirac to linearize the Klein–Gordon equation are in fact generators of a particular Clifford algebra [SOM 15].

Constructing universal Clifford algebra over R enables bidimensional algebra of complex numbers to be obtained. By considering the Cauchy–Riemann operator of the theory of holomorphic functions, and by multiplication with its complex conjugate, a two-dimensional Laplace is obtained. By means of Clifford algebra, this parameter can be generalized naturally to a greater dimension by introducing a generalized Cauchy–Riemann operator containing a scalar and avector with m dimensions. This operator's null solutions are generalized holomorphic functions called monogenic functions. We will tackle this fact in detail in monogenic signals and their application in signal processing.

An artificial neural network is a mathematical model inspired by biological neural networks that has been recognized as a useful tool in machine learning. For example, neural networks can learn non-linear relationships using input and output data from a number of non-linear calculation elements called neurons [HIT 13]. Neural networks can be extended to Clifford numbers called Clifford neural networks. Currently, Clifford neural networks are focused on multilayer types. Pearson was the first to propose a Clifford neural model, in which he formulated a model Clifford multilayer neural network, he also determined the learning algorithm by Clifford back-propagation and he thus clarified the approximation functions [PEA 94, PEA 01]. The generalization of the Fourier–Clifford transforms and quaternions to the Fourier–Clifford transforms with two sides is given and it enables linearity to be studied as far as convolution [HIT 14].

Quaternion Fourier transforms, Fourier–Clifford transforms, and Fourier transforms in geometric algebra have proved their usefulness as tools in processing non-marginal color images, image diffusion, multi-channel processing, the fast-matching pattern vector, quantum mechanics, proper complex signal non-stationarity and the spectral representation of analysis by Clifford wavelets. For information, we will cite the following references which explain the assertion above [GAL 13, LAS 93, LOU 99, MEN 13, EST 86, BAU 96, HAL 00] and [BAU 01].

4.6. Spinors and quantum signals

The Fourier transforms have undergone a spectacular development in the quantum domain.

A spinor is an element of a complex vector space. Spinors were introduced by Cartan in 1913. Subsequently, they have been used in quantum mechanics: the wave function of a fermion is represented by a Dirac bispinor. For particles with spin ½, this is expressed by the Dirac equation, for hypothetical particles with spin -3/2, it is the Rarita–Schwinger equation [RAR 41]. An exhaustive work has also been produced by Hirvensalo who developed the mathematical aspect for quantum information processing [HIR 13]. Visualization of the quantum Fourier transform has been achieved by Karafyllidis using a quantum simulator [KAR 03].

The main works on quantum signal processing and its counterpart information processing have been produced by [BEN 00, BET 03] and [NIE10]. In fact, in information processing as in physics, conventional, everyday observation produces an incomplete approximation explaining quantum reality. Quantum effects such as interference and entanglement do not play a direct role in conventional information processing, but they can be useful in practice for code breaking, creating unbreakable codes and greatly increasing calculation speeds.

5

The Contribution of Arithmetic to Signal Processing

The contribution of arithmetic to computers in general and to signal processing in particular can be appreciated from Inria's 1999 activity report. The project Arénaire thus details the contributions of arithmetic by giving the scientific foundations and applications for FPGA. The approach used is mainly appropriate to the use of arithmetic in hardware (integrated circuits or FPGA), digital calculation (floating point) and very precise calculations [MUL 99].

The basic arithmetical operation that a signal processor should carry out is fast multiplication in the form of a "product-sum" operation for time processing and a "complex multiplication" operation for the frequency domain and the FFT. These operations always involve structured data blocks taken from sensors.

As for us, we see the problem differently. We will focus our attention on the Fourier transform [GER 83, KAZ 03, GAO 14]. It can be noted from the last three references that Z's finite fields and subgroups are essential to establishing the formalism. Gauss' sums are also linked to other types of sums, in particular we will cite [BER 98], for the existing link between Gauss and Jacobi sums.

5.1. Gauss sums

A Gauss sum is a complex number whose definition uses harmonic analysis tools on a finite abelian group on the finite body Z/pZ where p represents an odd prime number and Z the group of relative integers. Gauss sums were introduced by Gauss in 1801 [GAU 66].

5.1.1. *Definition*

If ψ is a character of the additive group (**Fp**, +) and χ a character of the multiplicative group (**Fp***,), the Gauss sum associated with χ and ψ is the complex number G (χ, ψ) defined by:

$$G(\chi,\psi)=\sum_{x\in F_p^*}\chi(x)\psi(x) \qquad [5.1]$$

In terms of the Fourier transform, the application that links χ to G(χ^{-1}, ψ) can be considered to be the Fourier transform of the extension of χ to $\mathbf{F}_p$ by the equality χ(0) = 0 and the application that links to G(χ^{-1}, ψ) as the Fourier transform of the restriction of ψ to $\mathbf{F}_p^*$.

What is the link between the Gauss sum and the Fourier transform? The Fourier transform can be defined on the multiplicative group Fp^*:

$$\forall f\in C\left[F_p^*\right], F_{mul}(f):\begin{cases}\hat{F}_p^*\rightarrow C\\ \chi\rightarrow\sum_{x\in F_p^*}f(x)\chi(x)\end{cases} \qquad [5.2]$$

The Gauss sum can therefore be written as the multiplicative Fourier transform of an additive character, that is to say:

$$\forall\psi\hat{\in}\hat{F}_p,\forall\chi\in F_p^*,G(\chi,\psi)=F_{mul}(\psi)(\chi) \qquad [5.3]$$

Similarly, the Fourier transform on the additive group $\mathbf{F}_p$ can be defined. However, this requires the multiplicative characters $\chi\in\hat{F}_p^*$ to be extended according to $\hat{\chi}\in C\left[F_p\right]$, by asking $\hat{\chi}(0)=0$. The additive transform is therefore defined as follows:

$$\forall f\in C\left[F_p\right], F_{add}(f):\begin{cases}\hat{F}_p\rightarrow C\\ \psi\rightarrow\sum_{x\in F_p}f(x)\psi(x)\end{cases} \qquad [5.4]$$

Another writing of the Gauss sum is therefore obtained:

$$\forall\psi\hat{\in}\hat{F}_p,\forall\chi\in F_p^*,G(\chi,\psi)=F_{add}(\hat{\chi})(\psi) \qquad [5.5]$$

This enables us to break a multiplicative character down into an additive Fourier series.

Evaluating the Gauss sum:

– Consider the following expression:

$$G = \sum_{m=1}^{q-1} \left(\frac{m}{q} \right) e_q(m) \tag{5.6}$$

Where q is a prime number other than 1. It can be proven that $G^2 = q$, if $q \equiv 1(\text{mod}4)$ and G^2=-q if $q \equiv 3(\text{mod}4)$, although this does not determine G completely. The calculation is the following:

we have:

$$G^2 = \sum_{m_1=1}^{q-1} \sum_{m_2=1}^{q-1} \left(\frac{m_1 m_2}{q} \right) e_q(m_1 + m_2) \tag{5.7}$$

By changing the summation variable in the inner sum from m^2 to n, where $m_2 \equiv m_1 n(\text{mod } q)$, we obtain:

$$G^2 = \sum_{m_1=1}^{q-1} \sum_{n=1}^{q-1} \left(\frac{n}{q} \right) e_q(m_1 + m_1 n) \tag{5.8}$$

Now, we interchange the summation order, noting that:

$$\sum_{m_1=1}^{q-1} e_q\left[m_1(n+1)\right] = \begin{cases} q-1 & if, n \equiv -1(\text{mod } q) \\ -1 & \text{otherwise} \end{cases} \tag{5.9}$$

So:

$$G^2 = q\left(\frac{-1}{q} \right) + \sum_{n=1}^{q-1} \left(\frac{n}{q} \right)(-1) \tag{5.10}$$

$$G^2 = q\left(\frac{-1}{q} \right) = \begin{cases} q & si, q \equiv 1(\text{mod } 4) \\ -q & si, q \equiv 3(\text{mod } 4) \end{cases} \tag{5.11}$$

5.2. Applications for Gauss sums

Korobow gives the general framework of applications for Gauss sums with the necessary theoretical details [KOR 92].

Presently we know of several types of application for Gauss sums:

– signal processing;

– coding;

– programming technology;

– communication theory;

– factoring numbers;

– quantum entanglement;

– optical and photonic applications.

In signal processing, two works have held our attention; one concerns distributions [ELT 05], the other, correlation [SOS 06].

In coding, Hurt's "review" article draws together all the elements likely to apply Gauss sums to coding [HUR 97].

In *The Art of Computer Programming*, Knuth studies all the application possibilities for Gauss sums [KNU 89]. In terms of symbolic computation, we highlight Mihailescu and Vuletescu's work [MIH 10].

In terms of communication theory, a technical report established by Patterson defines the application methods for Gauss sums in communication [PAT 99].

Factoring numbers and quantum entanglement: theoretically, the literature is quite rich in publications in these areas, however, for the experimental aspect, developments are relatively extensive.

Factoring by Gauss sums has been addressed by Jones [JON 08], and by Mahesh *et al.* [MAH 07], as well as by Gilowski *et al.* [GIL 08]. The first work includes implementation techniques for nuclear magnetic resonance. The author considers a pendulum with two degrees of freedom; its excitation can be described by an amplitude and a direction angle in the form of a complex number. The individual terms of the Gauss sum are made by a resonant moment transferred to the pendulum. In the second work, the Gauss sum's factorization algorithm is implemented by the internal state of a Ramsey interferometer using cold atoms. The sequence of light pulses interacts with a group of cold rubidium atoms. The final population in the atomic levels involved determines a Gauss sum.

Gheorgiu [GHE 10] has demonstrated the construction of entangled bases in arbitrary dimensions via Gauss sums.

The factorization of Gauss sums and quantum phase measurement in a superconducting circuit has been studied by Ng and Nori [NG 10]. Two qubits of the Josephson phase are coupled to a photon field in a resonator. They show that the relative phase of the superposition of 2 Fock states can be imprinted in one of the 2 qubits. The qubit can thus be used to test and record the quantum coherence of 2 Fock states distinguishable for a monomode photon field inside a resonator. They show that the phase measuring method can be used to factorize Gauss sums of numbers higher than or equal to ten thousand in a product of prime numbers. The size of the factorizable number is mainly limited by the coherence time. If the resonator's relaxation time is between 10 μs and 1 ms, the order of the factorizable number can be higher than 10^7 N, where N is the number of photons in the resonator.

Several algorithms for factorizing large numbers based on physics have recently been introduced. Among them an algorithm developed by Mahesh *et al.* [MAH07]. uses Gauss sums to distinguish factors from non-factors. Two RMN techniques (nuclear magnetic resonance) that evaluate Gauss sums have been suggested. The first is based on differential excitation of a single spin's magnetization by a cascade of radiofrequency pulses. The second hinges on the spatial average and the selective refocalizing of magnetization by Gauss sums corresponding to these factors. All the factors of 16637 and 52882363 have been successively obtained [MAH 07].

On quantum entanglement, we will cite Michel Planat's article [PLA 05].

Matsutani has applied Gauss sums in the domain of optics with reasonably promising results [MAT 08].

Antonosyan *et al.* use Gauss sums for phase adaptation in the production of biphotons [ANT 12].

There are likely to be links between Gauss sums, Klusterman sums, monodromy groups and geometry, for which we can refer to the work of Katz [KAT 88].

6

Riemannian Geometry and Signal Processing

6.1. Context

Riemannian geometry admits as preliminary data a variety M whose points $x \in M$ are locally parametered by a finite number of real coordinates $x\mu$, and the metric given by the element of infinitesimal length:

– $ds^2 = g\mu\nu dx\mu dx\nu$

The distance between two points x, y∈M is given by:

– $d(x,y) = \text{InfLength}\gamma$

Where γ varies among the arcs joining x to y, and:

$$length\, \gamma = \int_x^y ds$$

The operational formalism of quantum mechanics joined to the analysis of logarithmic divergences of the trace of operators gives the generalization of the integro-differential calculus sought [CON 97].

The data of a geometric space is that of a spectral triple:

(A, H, D)

Where A is an involution algebra of operators on the Hilbert space H, and D an unbounded self-adjoint operator in H. The involution algebra A corresponds to the data in space M according to the classic Space ↔ Algebra duality in algebraic geometry. The operator $D^{-1} = ds$ corresponds to the infinitesimal element of length in Riemannian geometry. Riemannian geometry expands on methods from analytical geometry by using local coordinates to study curved spaces on which there are notions of angle and length. It introduces the founding concept of geometric variety. This is a direct generalization from Gauss's differential geometry of surfaces with n dimensions. It is a non-Euclidian geometry [HAT 02, JON 87, JOS 05, COO 08].

It enables shortest path problems to be tackled, in particular geodetics with their local and global aspects.

We will consider Riemannian geometry in four very specific cases:

– Voronoi diagrams, which have applications in mobile communications;

– two-dimensional signal (image) processing;

– Voronoi diagrams in geosciences;

– in geometric modeling for quantum communications.

6.1.1. *Voronoi diagrams in mobile communications*

The model used for mobile cell phone networks is n hexagonal mesh. A more realistic model shown is based on the Voronoi diagram and Delaunay triangulation. In the Voronoi mobile cellular network, cell size is determined by two basic factors: geographical position and the base stations' coverage zone. This mobile cellular network can be seen as a triangular planar graph in which the triangles are not regular. In cellular networks, simple diffusion algorithms create a spanning tree of a graph [HEA 13, TAY 12].

6.1.2. *Voronoi diagrams in image processing*

A region segmentation method is shown on the basis of a Voronoi polygon partition. The Voronoi diagram is used in a division and fusion environment evolving dynamically in relation to the image information. This segmentation is used to design a solution to the problem of detecting events in multi-source imagery [NAR 82].

6.1.3. *Voronoi diagrams in geosciences*

Original methods for surface meshing and volume meshing based on using Voronoi diagrams are suggested. The foundations for these two methods are identical: (1) a barycentric Voronoi optimization is used to obtain a controlled number of good quality elements, and (2) combinatorial considerations enable the final mesh to be constructed locally, possibly by modifying the initial model. The surface mesh method is automatic and enables a model to be simplified at a given resolution. The volume meshing method's originality lies in the fact that the elements generated are of different types. Prisms and pyramids are used to fill the models' very thin zones whereas the rest of the model is filled with tetrahedra [PEL 14].

6.1.4. *Voronoi diagrams in quantum communications*

Nielsen [NIE 10] introduces Voronoi diagrams in quantum communications and allows it to be proven by equivalence that the quantum Voronoi diagram on the degenerated Bloch sphere of pure quantum states coincides with the ordinary Euclidian Voronoi diagram, without ignoring the fact that quantum divergence is not defined here. It is then shown how to calculate the capacity of the Holevo channel [HOL 11] of 1 qubit quantum states and to provide a practical approximation algorithm based on Bregman sets. Finally, it defines the quantum-sided centroids which give the practical upper bounds on the Holevo capacity in linear time.

6.2. Riemannian varieties

Riemannian varieties can be deducted from the Hellinger distance, which is a means of evaluating the distance between two points for a given probability distribution. This kind of tool is very suitable for target recognition in a radar system. For this, we take a couple of probability distributions with a single estimator; evaluating their Hellinger distance provides a metric for a Riemannian variety. This metric is standard; it implies that an estimator can always be found in such a way that it renders the distance minimal. Choosing the best distribution reduces the estimator's calculation time [FRA 11].

There is a link between the ordinary spatial configuration algorithm and Riemannian geometry in the context of the Brain Computer Interface. It has been shown that the spatial filter in the shared spatial configuration and the logarithmic variance are extraction functionalities that can be summarized by calculating the Riemannian distance in the matrices' space covariance. This enables several approximations to be made apparent with respect to the space's topology [BAR 10].

6.3. Voronoi cells

In mathematics, a Voronoi diagram represents a particular decomposition of a metric space determined by the distances to a discreet set of the space's objects; this is a discreet set of points.

Voronoi diagrams are used for detection in wireless communications systems.

6.3.1. *Definition*

Take the affine plane $\mathbf{R}^2$. If S is a finite set of n points on the plane, the elements of S are called centers or sites.

The set of points that are closer to p than any other point of S is called the Voronoi region or Voronoi cell linked to an element, *p*, of *S*:

$$Vor_s(p) = \{x \in R^2 / \forall q \in S \|x - p\| \leq \|x - q\| \}$$

where $\|x - p\|$ designates the distance between x and p.

If we call the half-plane delimited by the segment mediator [pq], H(p,q) we have:

$$H(p,q) = \{x \in R^2 / \|x - p\| \leq \|x - q\|\}$$

$$Vor_s(p) = \bigcap_{q \in S/\{p\}} H(p,q)$$

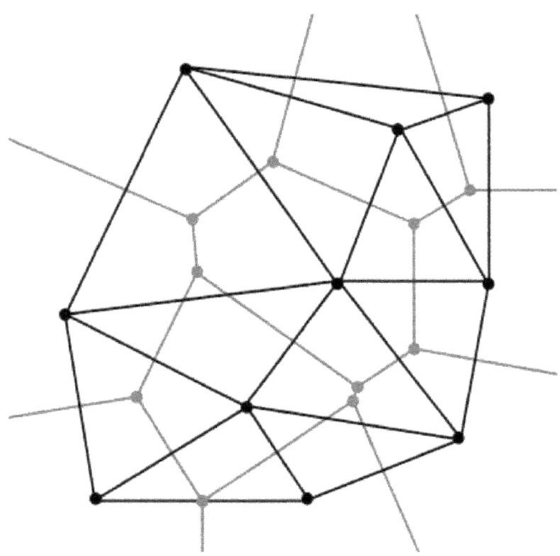

Figure 6.1. *Superposition of a Voronoi diagram (in red) and its Delaunay triangulation (in black). For a color version of this figure, see www.iste.co.uk/benslama/question.zip*

The Voronoi diagram of a discreet set S of points is the dual graph of the Delaunay triangulation [DEL 34], associated with S;

Each point of the Voronoi diagram constitutes a vertex in the Delaunay triangulation [LEE 80].

These vertices are linked to one another by a peak if and only if the cells are adjacent.

$$DEL(S) = \{(p,q) \in S^2 / Vor_s(p)Vor_s(q) \neq \varphi\} \qquad [6.1]$$

John Burkardt has suggested a file with four types of recording to represent a Voronoi diagram computationally [BUR 06]:

– *sxy*: point of set S, of co-ordinates (x, y);

– *vxy*: vertex of a Voronoi polygon of coordinates (x, y);

– *labc*: the right (carrying a polygon peak), of equation ax + by = c.

– *elv1v1*: peak of a polygon described by index 1 of its straight line and indices v_1 and v_2: of the vertices that form its extremities; if an index is equal to -1, this means that the vertex is infinite (half straight line, straight line if the vertices are at -1).

The following algorithms enable Voronoi diagrams to be obtained; we will cite them in order:

– Green and Sibson algorithm [GRE 78]

This algorithm enables Voronoi cells to be obtained from Voronoi or Thiessen polygons. A finite set of distinct points divides the plane into polygonal regions, each region containing one of the points and comprising the part of the plane closer to its point of definition than any other. The resulting planar subdivision is called Dirichlet tessellation; it is one of the most useful constructions associated with such a configuration of points. The regions, which we call tiles, are also known as Voronoi polygonsor Thiessen polygons [BRA 79]. We describe a recursive algorithm to calculate the tessellation very efficiently, and to discuss the problems that arise when it is implemented. Graphical output samples demonstrate the program's application on a modest scale; its efficiency enables it to be applied to large data sets, and a detailed discussion of space and time considerations is given, based in part on theoretical predictions and in part on the test function on up to 10,000 points.

– Shamos and Hoey algorithm [SHA 75]

This algorithm uses graph technology. A number of apparently unrelated problems involving the proximity of N points in the plane are studied, as well as how to find: a Euclidean minimum spanning tree, the smallest circle surrounding the set k, closer and more distant neighbors, the two closest points and a good straight triangulation of the line. For most of the problems considered O (n log n) is shown as the lower bound, and for all of these the upper limit best known is $O(N^2)$ or worse. The aim of this article is to present a single geometric structure, called the Voronoi diagram, which can be built rapidly and contains all the proximity information pertinent in a linear space. The Voronoi diagram is used to obtain O (n log n) algorithms for all the problems.

– Fortune algorithm [FOR 87]

This algorithm is based on a geometric transformation and a sweepline technique. We introduce a geometric transformation that enables Voronoi diagrams to be calculated using a sweepline technique. The transformation is used to obtain simple algorithms for calculating the Voronoi diagram of dot locations, line segment sites and the sites of weighed points.

6.4. Applications to Voronoi cells

Recent applications in stochastic geometry can be drawn on to analyze cellular systems. Stochastic geometry can be used to develop a mathematical work framework with the aim of modeling interference in networks; jammer locations are distributed according to a point process, often called a homogeneous Poisson point process. Analyzing downlink connection performances in a cell of fixed size, which is inscribed in a weighted Voronoi cell in a Poisson field of jammers, is very feasible [HEA 13]. In industry, cell tower locations have been modeled primarily by using a hexagonal deterministic grill. The real deployment uses are rarely regular; the same spacing between the nodes in the networks is a constant domain of Voronoi cells which makes the hexagonal grill unrealistic. Because of its relation with coverage zones, Taylor *et al.* suggest that the distribution of the Voronoi cell zone can be used as a test statistic in the general spatial modeling of cellular networks [TAY 12], [GHO 94].

By using tools from stochastic geometry, we define the user discovery performance of small cell access points, which enables us to draw from this the uplink capacity of small cells situated in the Voronoi cell of a macro cell base station representing the uncertainties linked to the random position, the density, the activated user, propagation channel and network interferences generated by uncoordinated activity [WIL 13]. It is also possible to obtain the mobile user's coverage probability and spectral efficiency in the worst cases by using stochastic geometry [JUN 13].

Energy efficiency gained importance when operational costs became a burden for ISP following the increasing demand made by data traffic in cellular networks. To this end, an energy efficiency model for cellular networks in Poisson–Voronoi tessellation has been proposed, hinging on spatial distributions of traffic load and energy consumption [XIA 13, BAN 12].

Algorithms for self-positioning autonomous mobile nodes are based on Voronoi tessellation [ZOU 13].

Pattern recognition and computer vision problems have been tackled by differential geometry. In this way, geometric representations and image laws require mathematical constraints to calculate traits and characteristics. In most cases, these constraints place the models on well-defined geometric spaces known as varieties. In such cases, the use of geometry for these varieties is important for understanding these objects and the semantic imaging attached to them. Statistics and algorithms take account of these varieties, enabling performance in visual application to be increased [LI 14, BAR 10], and [FAU 93].

PART 2

Applications

7

MIMO Systems

We will examine the principles of multiple input, multiple output systems with multi-carrier modulation OFDM (MIMO-OFDM). Since our objective is to study channel estimation methods for OFDM systems, it is essential to acquire a good understanding of OFDM systems before proceeding to investigate channel estimation methods

7.1. Introduction

OFDM is a method of coding digital signals by distribution at orthogonal frequencies in the form of multiple sub-carriers. It is based on multi-carrier communication technologies. The idea of multi-carrier communication consists of dividing the signal's total bandwidth into a number of sub-carriers; and the information is transmitted on each of the sub-carriers. Unlike the conventional multi-carrier communication method, in which the spectrum of each of the sub-carriers is non-overlapping and pass-band filtering is used to extract the frequency in question; the frequency spacing between the OFDM's sub-carriers is chosen so that the sub-carriers are mathematically orthogonal to one another. Sub-carrier spectrums overlap one another, but each individual sub-carrier can be extracted using a processing baseband. This overlapping property renders the OFDM more spectrally effective than the conventional multi-carrier communication method.

In the classic approach, the data traffic is applied directly to the modulator with a carrier frequency at the center of the transmission band f_0, f_{n-1}; that is to $(f_{n-1} + f_0)/2$, and the modulated signal occupies the entire width of band W. When the data are applied sequentially, a deep fading in the mobile channel can trigger burst errors. Figure 7.1 shows the transmission of symbols $S_0, S_1,..., S_{N-1}$, in series, where the shaded block indicates the position of the burst error, which only affects $K<N$ symbols.

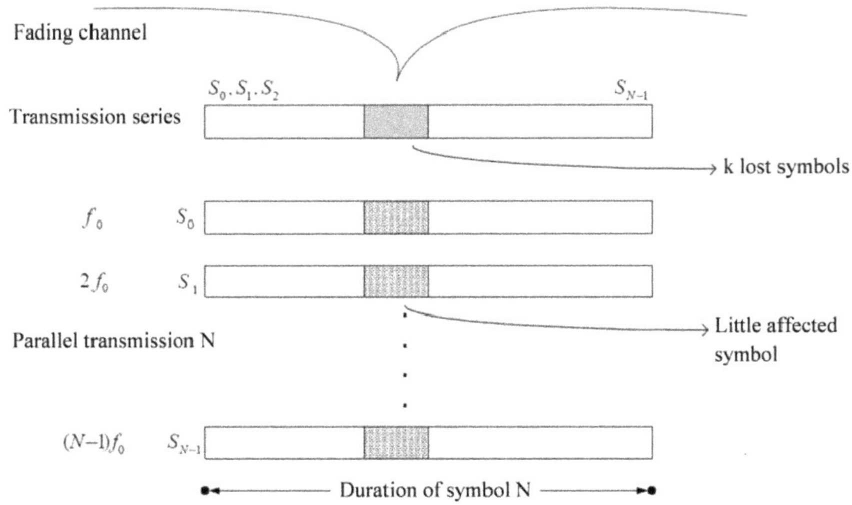

Figure 7.1. *Effect of a fading on serial and parallel systems*

During a symbol N's period of the conventional system, each OFDM modulator carries a symbol, and burst error causes significant breakdown in the duration of k symbols. However, if the burst error represents only a small fraction of the symbol's period, then each OFDM symbol will be affected slightly by the fading and can be demodulated correctly. Consequently, when the conventional system shows a burst error, few or no errors are produced using the OFDM approach.

Since the symbol's period has been increased, the channel's propagation delay represents a shorter fraction of the symbol's period than that of a conventional system, which makes the OFDM system less sensitive to ISI; which is another advantage of OFDM.

A drawback of the OFDM approach is its increased complexity compared to the conventional system. This complexity is due to the use of N modulators and filters at the transmitter and N demodulators and filters at the receiver. However, this complexity can be eliminated by using the FFT and the IFFT at the receiver and transmitter respectively, [BAR 91]. This is also represented in Figure 7.2.

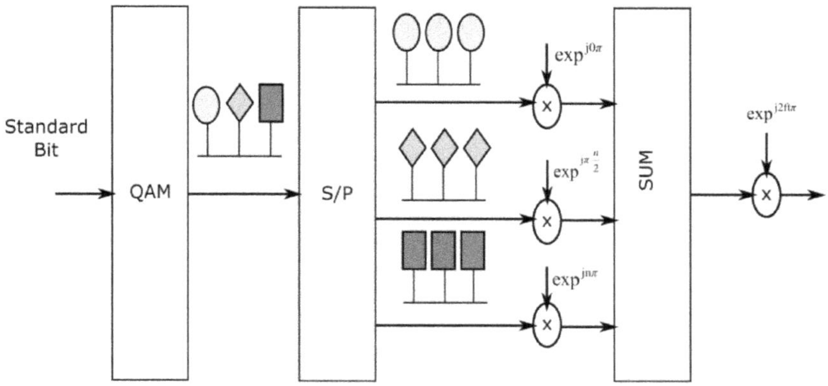

Figure 7.2. *An OFDM sub-carrier transmitter*

7.2. A brief history of OFDM

The concept of using transmission in parallel with data by means of frequency division multiplexing (FDM) was published in the 1960s [WEI 71]. In the United States, a diagram was made and published in January 1970. The idea was to use parallel data schemes and FDM with recovery sub-channels to avoid using high-speed equalization and to combat impulse noise and multi-path distortion as well as to use the entire bandwidth available. This was initially applied in military communications. In the domain of telecommunications, the terms multi-tone, multi-channel modulation and multi-carrier modulations (MCM) are in common use and sometimes they have been interchangeable with OFDM. In OFDM, each carrier is orthogonal to all the other carriers. This condition is not always maintained in MCM. OFDM is therefore an optimal version of multi-carrier transmission systems [RAE 99].

For a large number of sub-channels, the rows of sinusoidal generators and the coherent demodulator required in the parallel system become expensive and complex. The receiver needs to phase the demodulator carriers and sampling times precisely in order to maintain an acceptable interference between the sub-channels.

Weinstein and Ebert [WEI 71] have applied the discreet Fourier transform to parallel data as part of the modulation and demodulation process. But the problem with the FFT is that one can only use multiple frequencies of 1/T, where T is the symbol's period. In the 1980s, OFDM was studied for high-speed modems, digital mobile communications [POL 84] and high-density recording. In the 90s, OFDM was used for broad-band communications on mobile FM radio channels, digital broadband subscriber lines, asymmetrical digital subscriber lines, high speed digital subscriber lines, digital audio diffusion (DAB) and analog HDTV.

7.3. Multi-carrier technology

The principle of multi-carrier transmission is to convert a high-rate serial data stream to a reduced-rate parallel multiple secondary flow. Each secondary flow is modulated on another sub-carrier. Since the symbol rate on each sub-carrier is much lower than the initial information symbol rate (in series), the effects of the propagation delay; that is the ISI, decrease significantly, while still reducing the equalizer's complexity.

In multi-carrier modulation the available bandwidth W is divided into N sub-bands or sub-carriers, each of width $\Delta f = W/N$. This subdivision is illustrated in Figure 7.3.

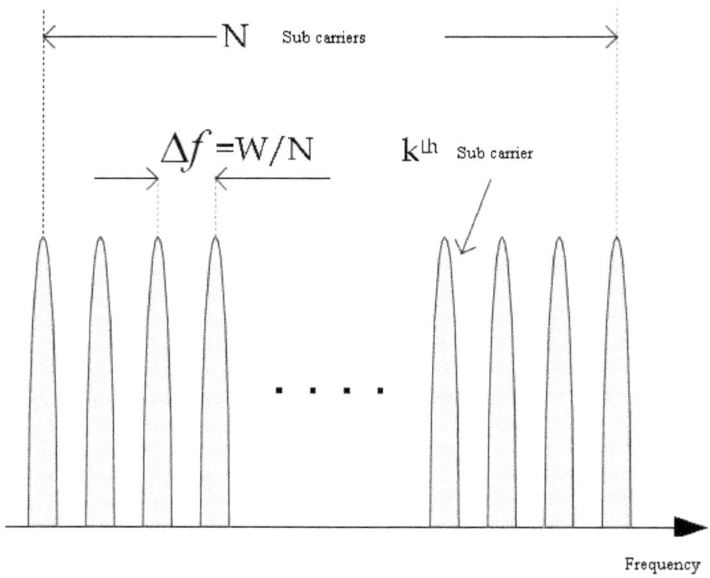

Figure 7.3. *Subdivision of the bandwidth into N sub-carriers*

Instead of transmitting the data-symbols in series, the multi-carrier transmitter divides the data into blocks of N data symbols that are transmitted in parallel by modulating the N carriers. A symbol's duration for a modulated carrier is $Tc = 1/\text{W}$.

The multi-carrier signal can be written as a series of modulated carriers [PRO 95]:

$$s(t)=\sum_{k=0}^{N-1} x_k \psi(t) \qquad [7.1]$$

where:

– x_k is the data symbol modulating the k-th sub-carrier;

– $\psi_k(t)$ is the modulation wave of the k-th sub-carrier;

– s (t) is the modulated signal in multi-carriers.

The process of generating a modulated sub-carrier signal is illustrated in Figure 7.4.

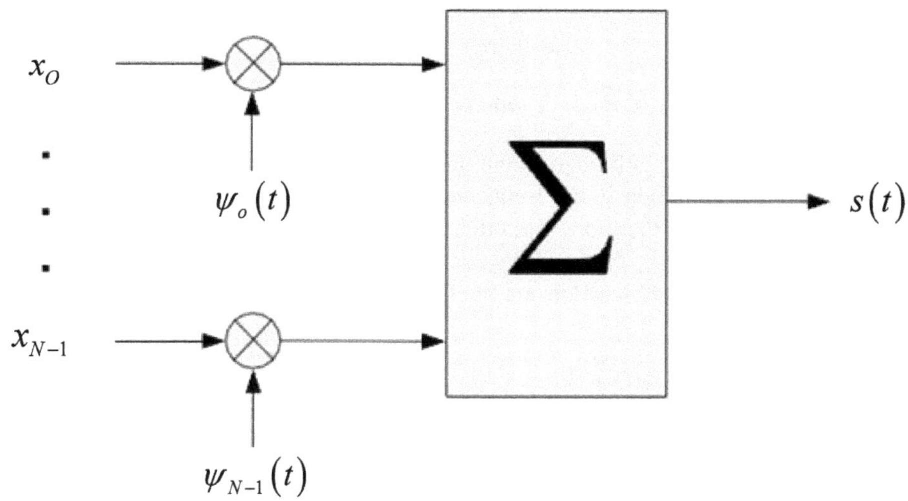

Figure 7.4. *Multi-carrier modulation*

A number of measures can be taken when designing a multi-carrier system to mitigate the effects of fading:

– In the time domain, the symbol's duration can be rendered much longer than the channel's maximum excess delay. This can be done by choosing $Ts >> \tau$ max;

– in the frequency domain, the sub-carriers' bandwidth can be reduced in comparison to the channel's coherence bandwidth Bcoh>>W/N. The sub-bands therefore undergo a flat fading, which amounts to equalizing a single complex carrier multiplication;

– it can be noted that the stages mentioned above are two aspects of the same phenomenon; everything depends on the time or frequency domain.

7.4. OFDM technique

Multi-carrier modulations that use orthogonal waves to modulate the sub-carriers are called OFDM systems. Since the sub-carriers are modulated by orthogonal waves, they have overlapping spectrums, which gives the spectrum greater efficiency.

A general set of orthogonal waves is given by:

$$\psi_k(t) = \begin{cases} \frac{1}{\sqrt{T_s}} e^{j2\pi f_k t} & t \in [0, T_s] \\ 0 & \text{elsewhere} \end{cases} \qquad [7.2]$$

where f_k is the frequency of the k-th sub-carrier, with $k = 0, 1, \ldots, N-1$

Since the wave $\psi_k(t)$ is limited in the time window [0, Ts], the k-th sub-carrier's spectrum is a sinc function in the frequency domain as illustrated in Figure 7.5 for $N = 3$ and $N = 64$. Furthermore, although the spectrum is overlapped, it does not cause any interference in the carrier's location, which is due to the sub-carriers' orthogonal nature. The sub-carriers are in effect orthogonal, since:

$$\int_0^{T_s} \psi_k(t).\psi_1^*(t)dt = \delta[k-1] \qquad [7.3]$$

where

– $\psi^*(t)$ is the wave's complex conjugate function $\psi(t)$, and

$$\delta[k-1] = \begin{cases} 1 & k = 1 \\ 0 & k \neq 1 \end{cases}$$

is the Kronecker delta function

Figure 7.5(a) also illustrates the significance of frequency synchronization in OFDM systems. The spectrum of the other carriers is only different from zero in the carriers' exact locations, and the orthogonality is lost in the absence of the carriers' synchronization. The Gibbs phenomenon is evident looking at the edges of the spectrum for $N = 64$ in Figure 7.5(b).

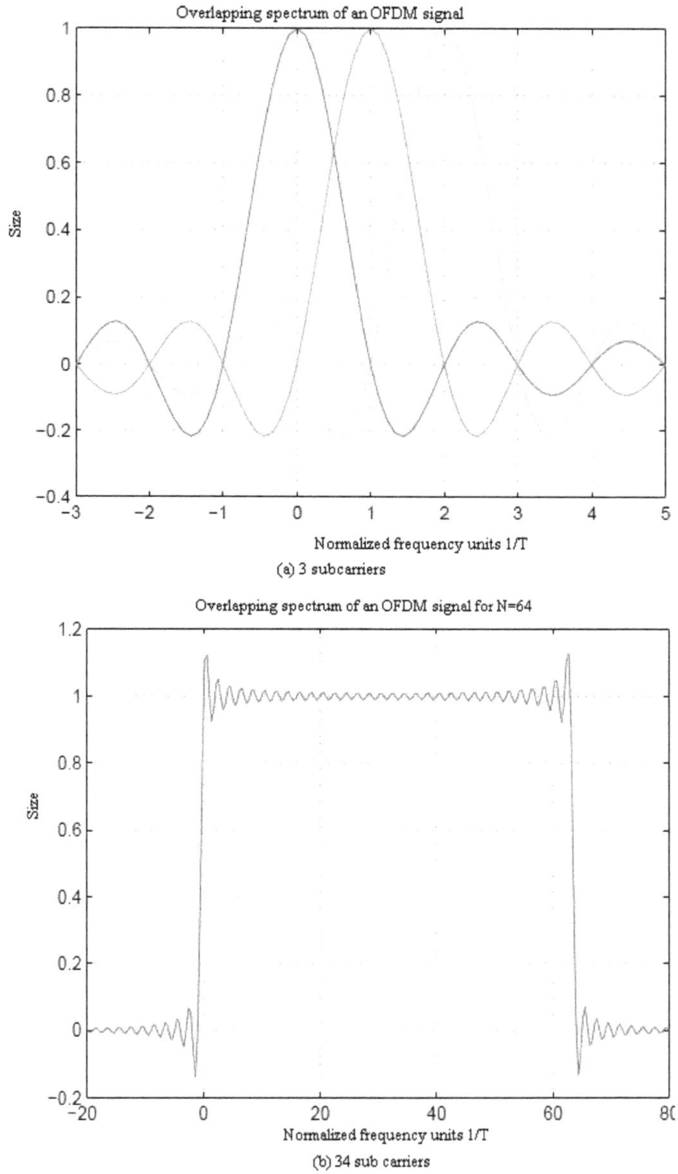

Figure 7.5. *The overlapped spectrum of an OFDM signal. For a color version of this figure, see www.iste.co.uk/benslama/quantum.zip*

Demodulation in OFDM systems is based on the sub-carriers' orthogonality and is composed of a series of N assorted filters. By implementing assorted filter correlators, we can write [VAL 14]:

$$y_k = \int_0^{T_s} s\,(t).\psi_k^*(t)dt \qquad [7.4]$$

where:

– y_k is the k-th sub-carrier's demodulated symbol

– $s(t)$ is the signal transmitted as given in [7.1].

Demodulation in OFDM can be applied as shown in Figure 7.6. Implementing such a system becomes complex as the number of sub-carriers increases. It should be noted that by replacing the transmitter oscillators with the IDFT and the assorted filter with the DFT, there will be the same result [BHA 14].

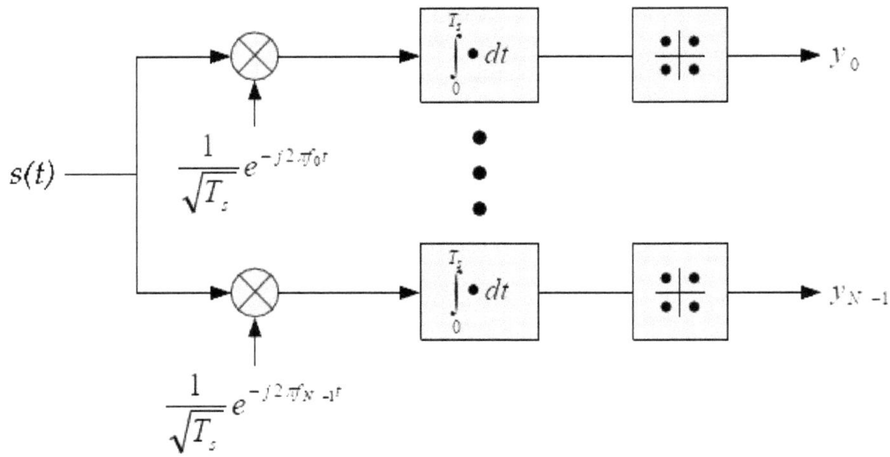

Figure 7.6. *OFDM demodulation using correlators*

7.5. Generating OFDM symbols

The diagram of a MIMO OFDM system is shown in Figure 7.7.

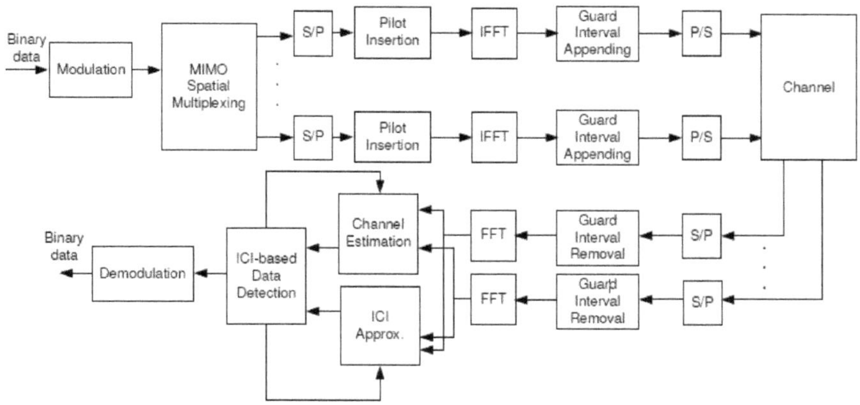

Figure 7.7. *Diagram of a MIMO OFDM system*

An OFDM symbol can be generated in the digital domain before modulation on a carrier for transmission. To generate an OFDM symbol, a serial digital data stream is first modulated using a known modulation technique such as phase-shift keying (PSK) or quadrature amplitude modulation (QAM). These data symbols are then converted into parallel streams before modulating the sub-carriers. The sub-carriers are sampled at the sampling rate N/Ts. The frequency separation between two adjacent sub-carriers is 2π/N. Finally, the samples on each sub-carrier are added together to form an OFDM sample. An OFDM symbol generated by an OFDM system of N sub-carriers is composed of N samples and the m-th sample of an OFDM symbol is given by:

$$x_m = \sum_{k=1}^{N-1} X_k e^{j\frac{2\pi mk}{N}} \qquad 0 \le m \le N-1 \qquad [7.5]$$

where X_k is the symbol transmitted on the k-th carrier.

Equation [7.5] is equivalent to the inverse discreet Fourier transform of N points (IDFT) on the data sequence with a graduation factor omitted. It is well-known that the IDFT can be implemented effectively by using the Inverse Fast Fourier Transform (IFFT). Consequently, in practice the IFFT is carried out on the data sequence at the OFDM transmitter for the baseband modulation and the FFT is carried out on an OFDM receiver for the baseband demodulation. The size of the FFT and IFFT is N, which is equal to the number of secondary channels available for transmission. However, all the channels should be active. The secondary channel's bandwidth is given by:

$$f_{sc} = \frac{1}{T_s} = \frac{f_{samp}}{N}$$

where f_{samp} is the sampling rate and Ts the symbol's time.

Finally, an OFDM baseband symbol is modulated by a carrier to become a passband signal and transmitted to the receiver. In the frequency domain, this corresponds to translating all the baseband subcarriers to the carrier frequency simultaneously.

7.6. Inter-symbol and inter-carrier interference

In a multi-path environment, a transmitted symbol takes different lengths of time to reach the receiver by different propagation paths. From the perspective of the receivers, the channel presents a propagation delay in which the received symbol's duration is extended. As a result of prolonging the symbol's duration, the symbol received is covered by the previous received symbols, which results in inter-symbol interference (ISI) [GUV 13]. In OFDM systems, ISI usually refers to the previous OFDM symbols interfering with the OFDM symbol.

In OFDM, the sub-carrier spectrums overlap, but always remain orthogonal to one another. This means that at the maximum of each spectrum of the subcarriers, all the spectrums of the other sub-carriers are equal to zero [LIN 00]. The receiver samples the data symbols on individual sub-carriers at the maximum points and demodulates them free of any interference with the other sub-carriers. The interference caused by the data symbols on the adjacent sub-carriers is called inter-carrier interference (ICI).

The sub-carriers' orthogonality is apparent in the time domain or in the frequency domain. From the point of view of the time domain, each sub-carrier is a sinusoid with a whole number of cycles in an FFT interval. From the point of view of the frequency domain, this corresponds to each sub-carrier having the maximum value at its own central frequency and zero at each of the other sub-carriers' central frequency.

A sub-carrier's orthogonality in relation to the other sub-carriers is lost if the sub-carrier has a non–null spectral value at the other sub-carriers' frequencies. In the time domain, this means that the corresponding sinusoid no longer has a whole number of cycles in the FFT interval.

ICI occurs when the multi-path channel varies in an OFDM symbol's duration [VAL 14]. When this happens, the Doppler effects on each multi-path involve a frequency shift on the sub-carriers, the result being the loss of orthogonality between them. In the time domain, this means that the whole number of cycles for each sub-carrier in the current symbol's FFT interval is no longer maintained because of the phase transition introduced by the previous symbol. In conclusion, any shift between the transmitter's and receiver's sub-carrier frequencies also present ICI to an OFDM symbol.

7.7. Cyclic prefix

The OFDM is resistant to ISIs, because its symbol duration is long compared to those of serial data symbols. For an OFDM transmitter of N sub-carriers, if a data symbol's duration is T', an OFDM symbol's duration on leaving the transmitter:

$$Ts = T'N \quad [7.6]$$

Thus, if a multi-path channel's propagation delay is greater than T but less than Ts, the data symbol in the serial data stream undergoes a frequency selective fading, whereas the data symbol on each sub-carrier only undergoes a flat fading. To eliminate ISI almost entirely, a cyclic prefix (CP) is introduced for each OFDM symbol, as shown in Figure 7.8, and the length of the CP chosen should be longer than the propagation delay encountered. Moreover, the CP can maintain orthogonality between the sub-carriers, which implies that the ICI is null.

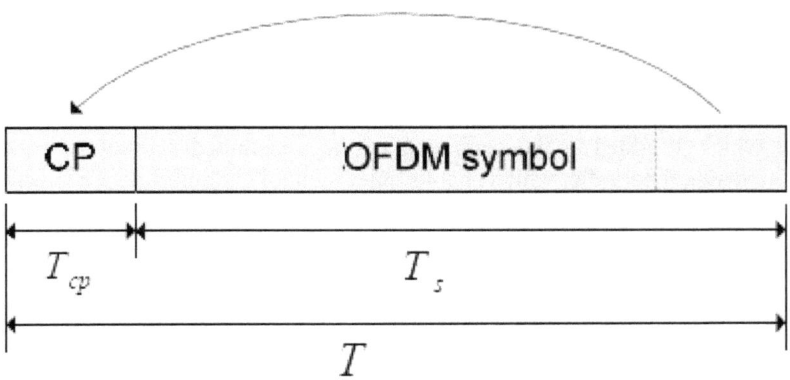

Figure 7.8. *Cyclic prefix*

The cyclic prefix's length should be at least equal to the significant part of the channel's impulse response encountered by the transmitted signal. Thus, the cyclic prefix shows the following advantages:

– ICI is avoided since the cyclic prefix plays the role of space guard between successive symbols;

– the linear convolution with the channel's impulse response is converted into a cyclic convolution. As the cyclic convolution in the time domain corresponds to a scalar multiplication in the frequency domain, the sub-carriers become orthogonal, and the ICI will be null.

The CP's advantages are not without cost. The energy transmitted, required to transmit the signal, increases with the length of the cyclic prefix. This SNR loss is given by:

$$SNR_{loss} = -10\log_{10}\left(1-\frac{T_{cp}}{T}\right) \quad [7.7]$$

Where: Tcp is the length of the cyclic prefix, Ts is the symbol's time, and T = Tcp + Ts is the length of the symbol transmitted.

7.8. Mathematical model of the OFDM system

The basic idea of OFDM is to divide the available spectrum into several sub-channels (sub-carriers) by making all the narrowband sub-channels then undergo a flat fading, which facilitates or removes the need for equalization. For greater spectral efficiency, the sub-channels' frequency response is overlapped and orthogonal, hence the name OFDM. This orthogonality can be maintained completely even if the signal crosses a time-dispersive channel by introducing a cyclic prefix [GER 05]. The cyclic prefix is a copy of the last part of the OFDM symbol which is added at the beginning of the transmitted symbol, to make the transmitted signal periodic, which enables inter-symbol and inter-carrier interference to be avoided [GER 05]. Although the cyclic prefix shows a loss in the signal-noise ratio (SNR), it is overall a small price to pay to attenuate interference.

A schematic diagram of an OFDM system is shown in Figure 7.9. For this system, we use the hypotheses:

1) a cyclic prefix is used;

2) the channel's impulse response is shorter than the cyclic prefix;

3) the transmitter and receiver are perfectly synchronized;

4) the fading is slow enough for the channel to be considered constant during an OFDM symbol interval.

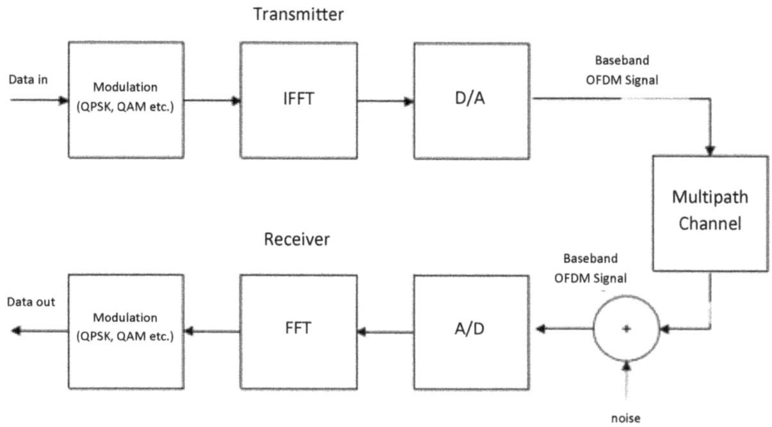

Figure 7.9. *A basic FFT OFDM transmitter-receiver [GER 05]*

In order to make an in-depth analysis of the system, we will put forward some hypotheses and develop a model for this system. OFDM systems are placed into two different classes, in continuous time and in discreet time.

7.8.1. *Continuous time model*

Previous OFDM systems did not use any digital modulation and demodulation, consequently the OFDM model shown below can be considered an ideal OFDM, which is digitally synthesized in practice. We begin with the wave forms used in the transmitter and we continue with the receiver.

7.8.1.1. *Transmitter*

Consider an OFDM system with N sub-carriers, a bandwidth of W hertz and symbol duration of T seconds, of which Tcp seconds represent the length of the cyclic prefix. The transmitter uses the following wave forms:

$$\Phi_k(t) = \begin{cases} \frac{1}{\sqrt{T-T_{cp}}} e^{j2\pi w \frac{k}{N} f_k (t-T_{cp})} & t \in [0, T\ ] \\ 0 & \text{elsewhere} \end{cases} \quad [7.8]$$

where

$$T = (N/W) + T_{cp}$$

it is noted that $\Phi k(t) = \Phi k(t+N/W)$, where T is in the cyclic prefix [0, T_{cp}].

Since Φk(t) is the modulated rectangular pulse on the carrier frequency kW/N, the common interpretation of OFDM is that it uses N sub-carriers, each of which carries a low bit rate. The baseband signal transmitted for the number, l, of OFDM symbols is given by:

$$s_l(t) = \sum_{k=0}^{N-1} x_k(l)\phi_k(t - lT) \qquad [7.9]$$

where x_0, x_1, …, x_{N-1}, are complex symbols and form a group of the signal's constellation points. If a sequence of OFDM symbols is emitted, the transmitter output will be a juxtaposition of individual OFDM symbols:

$$s(t) = \sum_{l=-\infty}^{+\infty} s_l(t) = \sum_{l=-\infty}^{+\infty} \sum_{k=0}^{N-1} x_{k,l}(l)\phi_k(t - lT) \qquad [7.10]$$

7.8.1.2. *Physical channel*

We suppose that the support for the physical channel's impulse response $g(\tau, t)$ (eventually variable in time) is limited in the interval τ [0, T_{cp}]; that is to the length of the cyclic prefix. The signal received becomes:

$$r(t) = g(\tau, t) * s(t) + n(t) = \int_0^{\tau_{cp}} g(\tau, t)s(t - \tau)d\tau + \tilde{n}(t) \qquad [7.11]$$

where $\tilde{n}(t)$ is the channel's complex additive white Gaussian noise.

7.8.1.3. *The receiver*

The OFDM receiver is made up of a filter bank adapted to the last part [T_{cp},T] of the waves from the transmitter Φk (t), that is to say,

$$\psi_k(t) = \begin{cases} \Phi^*(T - t) & t \in [0, T_{cp}] \\ 0 & \text{elsewhere} \end{cases} \qquad [7.12]$$

In fact, this means that the cyclic prefix is suppressed in the receiver. Since the cyclic prefix contains all of the previous symbol's ISIs, the sampled output from the set of receiving filters does not contain an ISI. Consequently we can ignore the time index l when calculating the sampled output at the k-th adapted filter. By using [7.10], [7.11] and [7.12], we will obtain:

$$y_k = \int_{\tau_{cp}}^{T} \left(\int_{0}^{\tau_{cp}} g(\tau,t) \left[\sum_{k=0}^{N-1} x_k \phi_k \left(t-\tau\right) \right] d\tau \right) \phi_k^*(t)dt + \int_{\tau_{cp}}^{T} \tilde{n}(T-t)\phi_k^*(t)dt \qquad [7.13]$$

We consider that the channel is fixed on the OFDM symbol's interval, and it is designated g(τ), which gives:

$$y_k = \sum_{k=0}^{N-1} x_k \int_{\tau_{cp}}^{T} \left(\int_{0}^{\tau_{cp}} g(\tau)\phi_k(t-\tau)d\tau \right) \phi_k^*(t)dt + \int_{\tau_{cp}}^{T} \tilde{n}(T-t)\phi_k^*(t)dt \qquad [7.14]$$

The integration intervals are $T_{cp}<t<T$ and $0<\tau<Tcp$, which means that $0<t-\tau<T$ and the inner integral can be written as follows:

$$\int_{0}^{T_{cp}} g(\tau)\phi_k(t-\tau)d\tau = \int_{0}^{T_{cp}} g(\tau)\frac{e^{j2\pi k(t-\tau-T_{cp})\frac{W}{N}}}{\sqrt{T-T_{cp}}}d\tau = \frac{e^{j2\pi k(t-T_{cp})\frac{W}{N}}}{\sqrt{T-T_{cp}}} \int_{0}^{T_{cp}} g(\tau)\frac{e^{-j2\pi k\tau\frac{W}{N}}}{\sqrt{T-T_{cp}}}d\tau$$

$T_{cp} \leq t \leq T$

The latter part of this expression is the frequency response sampled from the channel at frequency $f = K'W/N$, that is to say, the frequency of k'-sub-carrier:

$$h_{k'} = G(k'\frac{W}{N}) = \int_{0}^{T_{cp}} g(\tau)e^{-j2\pi k'\tau\frac{W}{N}}d\tau \qquad [7.15]$$

where G(f) is the Fourier transform of g(τ).

By using this notation, the output from the receiver's set of filters can be simplified to:

$$y_k = \sum_{k=0}^{N-1} x_{k'} \int_{T_{cp}}^{T} \frac{e^{j2\pi k'(t-T_{cp})\frac{W}{N}}}{\sqrt{T-T_{cp}}} h_{k'}\phi_k^*(t)dt + \int_{\tau_{cp}}^{T} \tilde{n}(T-t)\phi_k^*(t)dt = \sum_{k=0}^{N-1} x_{k'}h_{k'} \int_{T_{cp}}^{T} \phi_k^*(t)\phi_k(t)dt + n_k \qquad [7.16]$$

where:

$$n_k = \int_{T_{cp}}^{T} \tilde{n}(T-t)\phi_k^*(t)dt$$

Since the transmitter's receivers Φk(t) are orthogonal:

$$\int_{T_{cp}}^{T} \phi_{k'}^{*}(t)\phi_{k}(t)dt = \int_{T_{cp}}^{T} \frac{e^{j2\pi k'(t-T_{cp})\frac{W}{N}} e^{j2\pi k(t-T_{cp})\frac{W}{N}}}{\sqrt{T-T_{cp}}\sqrt{T-T_{cp}}} dt = \delta[k-k'] \qquad [7.17]$$

Where δ[k] is the Kronecker delta function.

Thus, [7.16] becomes:

$$Y_k = h_k x_k + n_k \qquad [7.18]$$

where nk is an additive white Gaussian noise.

The cyclic prefix has a double advantage: it simultaneously avoids the ISI (as it acts as a space guard) and the ICI (as it maintains the sub-carriers' orthogonality). By reintroducing the time index l, we can see the OFDM system as a set of parallel Gaussian channels according to Figure 7.10.

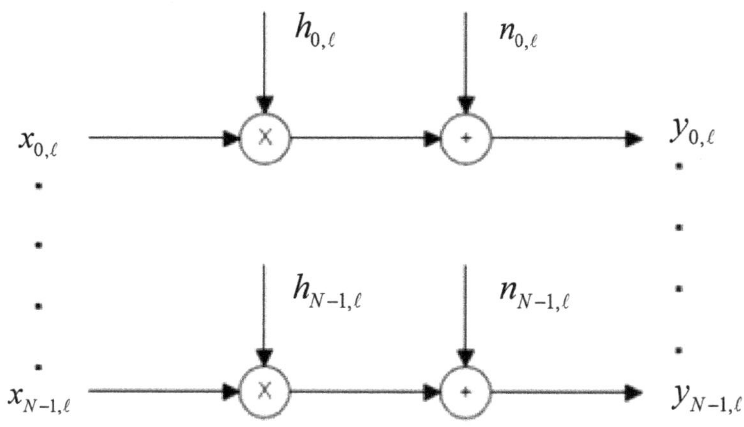

Figure 7.10. *The continuous time OFDM system interpreted as parallel Gaussian channels*

7.8.2. *Discrete time model*

In the discrete time model, modulation and demodulation are replaced by an inverse DFT (IDFT) and a DFT respectively, and the channel is a discrete-time convolution. The cyclic prefix functions in the same way as in the continuous-time model, and the calculations can be carried out in the same way. The main difference between the two models is that all the integrals are replaced by sums.

Using a cyclic prefix longer than the channel's impulse response transforms the linear convolution into a cyclic convolution.

The whole OFDM system can be written as follows:

$$y_l = DFT(IDFT(x_l) \otimes g_l) + \tilde{n}_l = DFT(IDFT(x_l) \otimes g_l) + n_l \quad [7.19]$$

where $\otimes$ indicates the cyclic convolution, and y_l contains N received data points, x_l contains N transmitted constellation points, gl the channel's impulse response (filled with zeros to obtain the length N), and $\check{n}_l$ is the channel's noise. As the channel's noise is supposed to be white and Gaussian, the term n_l= DFT ($\check{n}_l$) represents a non-Gaussian, non-correlated noise. Moreover, we make use of the fact that the DFT of two convoluted cyclic signals is equivalent to the product of their individual DFTs.

By designing the multiplication element by element by (.), the expression above can be written as:

$$Yl = x_l.DFT(g_l) + n_l = x_l.h_l + n_l \quad [7.20]$$

where h_l = DFT(g_l) is the channel's frequency response. Thus, we have obtained the same type of parallel Gaussian channels as for the continuous-time model. The only difference lies in the attenuations of the h_l channels that are given by the DFT of N points of the discrete-time channel, in place of the sampled frequency response given by [7.17].

7.9. MIMO channels

The current principle difficulty is that users demand higher flow rates for their applications despite the usable spectrum being limited (technically and legally). This is due to the growth in popularity of mobile applications such as cell phones and wireless internet access. Wireless systems do not have the option of simply adding an additional cable as wired systems or fiber optic based systems do. Consequently, spectral efficiency should be increased in order to permit a higher flow. But clients do not solely want rapid access to data; this access must also be reliable (service quality).

MIMO systems seem to be able to solve this problem, at least temporarily. Rather than transmitting a single signal into the "air" from transmitter to receiver, several independent signals are emitted on the shared "air" channel using multiple antennae to transmit and receive. The idea seems quite trivial, but sending signals in the same frequency band on a shared channel is not usually possible, because the signals interfere with one another and cannot be decoded easily at the receiver [VIT 95].

In most applications, each signal sent from a transmitting antenna reaches the receiving antenna via multiple paths. This phenomenon is called multi-path propagation; it is produced by electromagnetic waves that are reflected on walls and other objects. The signal arriving at a receiver is therefore a superposition of versions scaled and delayed from the original signal. Multi-path propagation is generally considered to be a nuisance as it distorts the signal, and usual systems try to circumvent it by establishing a line of sight connection (LOS) [VIT 95].

Wireless MIMO channels are characterized as variable in time. The transmitted signal arrives at the receiver from different directions on multiple paths because it meets obstacles and reflectors in the wireless propagation channels. The signal received is the sum of these waves, which arrive with a different phase, a different amplitude and a different delay. If multi-path signals' waves undergo a phase inversion, the signals' intensity at the receiver may be reduced.

This type of phenomenon is called multi-path fading, which is generally modelled as Rayleigh fading [CHA 05a, CHA 05b]. It is a major obstacle to these wireless channels' reliability; it causes significant degradation in the performance of digital wireless communications systems.

Different arrival times for the transmitted signal's multiple reflections can trigger ISI. This time, dispersion of the channels is called multi-path propagation delay or propagation channel delay, which is an important parameter for obtaining wireless systems' performance. Current multi-path methods are the multi-path propagation delay's root mean squared (RMS) and the maximal multi-path delay. For a reliable wireless communication, without using an adaptive equalizer or another anti-multi-path technique, the rate of data transmitted should be smaller than the coherent bandwidth, which is approximately equal to the inverse of the channel's multi-path propagation.

To find the channel's matrix, a test phase or known signals are transmitted by the transmitter. This diminishes the overall throughput as no information is transmitted during this phase. This loss is generally smaller than the additional capacity obtained by using a second flow.

7.10. The MIMO channel model

To simplify the writing of a MIMO system with T transmission antennae and R reception antennae, suppose that $T = R = 4$. A general configuration of this kind is shown in Figure 7.11. It is easy to change the number of transmission or reception antennae.

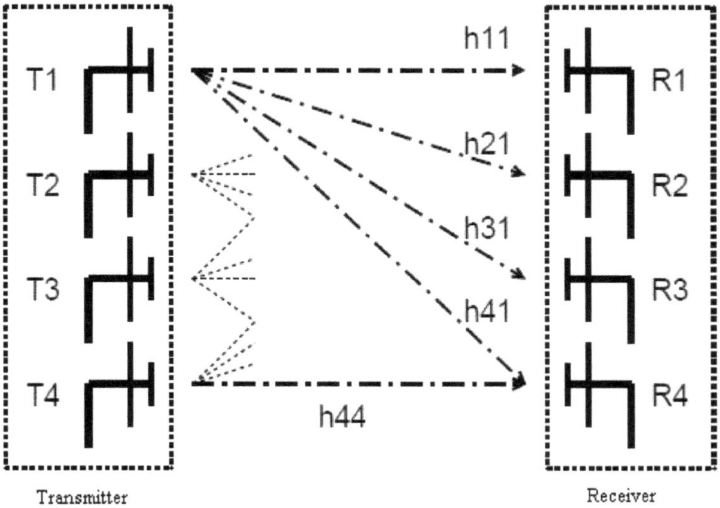

Figure 7.11. *A standard approach for a MIMO–OFDM system with four transmitting antennae and four receiving antennae*

The system shown in Figure 7.11 can be written in the frequency domain as follows:

$$\underbrace{\begin{bmatrix} y_r = 1 \\ y_r = 2 \\ y_r = 3 \\ y_r = 4 \end{bmatrix}}_{Y_{R\times 1}} = \underbrace{\begin{bmatrix} h_{r=1,\tau=1} & h_{r=1,\tau=2} & h_{r=1,\tau=3} & h_{r=1,\tau=4} \\ h_{r=2,\tau=1} & h_{r=2,\tau=2} & h_{r=2,\tau=3} & h_{r=2,\tau=4} \\ h_{r=3,\tau=1} & h_{r=3,\tau=2} & h_{r=3,\tau=3} & h_{r=3,\tau=4} \\ h_{r=4,\tau=1} & h_{r=4,\tau=2} & h_{r=4,\tau=3} & h_{r=4,\tau=4} \end{bmatrix}}_{H_{R\times T}} \cdot \underbrace{\begin{bmatrix} x_{\tau=1} \\ x_{\tau=2} \\ x_{\tau=3} \\ x_{\tau=4} \end{bmatrix}}_{X_{T\times 1}} + \underbrace{\begin{bmatrix} n_{r=1} \\ n_{r=2} \\ n_{r=3} \\ n_{r=4} \end{bmatrix}}_{N_{R\times 1}}$$

Suppose that only an ante antenna is currently transmitting the signal and none of the others are sending any signal. In this case, the equation is simplified to:

$$\underbrace{\begin{bmatrix} y_r = 1 \\ y_r = 2 \\ y_r = 3 \\ y_r = 4 \end{bmatrix}}_{Y_{R\times 1}} = \underbrace{\begin{bmatrix} h_{r=1,\tau=1} & h_{r=1,\tau=2} & h_{r=1,\tau=3} & h_{r=1,\tau=4} \\ h_{r=2,\tau=1} & h_{r=2,\tau=2} & h_{r=2,\tau=3} & h_{r=2,\tau=4} \\ h_{r=3,\tau=1} & h_{r=3,\tau=2} & h_{r=3,\tau=3} & h_{r=3,\tau=4} \\ h_{r=4,\tau=1} & h_{r=4,\tau=2} & h_{r=4,\tau=3} & h_{r=4,\tau=4} \end{bmatrix}}_{H_{R\times T}} \cdot \underbrace{\begin{bmatrix} x_{\tau=1} \\ 0 \\ 0 \\ 0 \end{bmatrix}}_{X_{T\times 1}} + \underbrace{\begin{bmatrix} n_{r=1} \\ n_{r=2} \\ n_{r=3} \\ n_{r=4} \end{bmatrix}}_{N_{R\times 1}}$$

This can be simplified to:

$$\underbrace{\begin{bmatrix} y_r = 1 \\ y_r = 2 \\ y_r = 3 \\ y_r = 4 \end{bmatrix}}_{Y_{R\times 1}} = \underbrace{\begin{bmatrix} h_{r=1,\tau=1} \\ h_{r=2,\tau=1} \\ h_{r=3,\tau=1} \\ h_{r=4,\tau=1} \end{bmatrix}}_{H_{R\times 1}} \cdot x_{\tau=1} + \underbrace{\begin{bmatrix} n_{r=1} \\ n_{r=2} \\ n_{r=3} \\ n_{r=4} \end{bmatrix}}_{N_{R\times 1}}$$

7.11. MIMO OFDM channel model

The different OFDM sub-channels in the SISO–OFDM system can be processed separately. This enables a simple model for a MIMO–OFDM system to be formulated; the whole system can be seen as a stack of N different MIMO systems. A graph showing a system of this kind is shown in Figure 7.12.

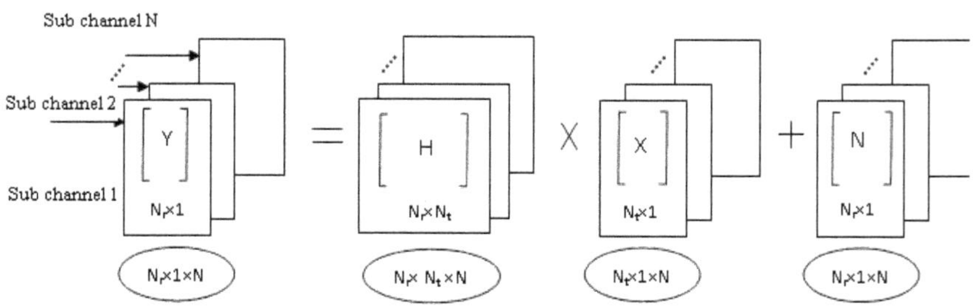

Figure 7.12. *Model of a MIMO–OFDM system*

8

Minimizing Interferences in DS–CDMA Systems

Many works have tackled minimization of interferences in CDMA systems using different techniques. We present an approach to studying convolutive codes and their processing using wavelets.

8.1. Convolutional encoding

8.1.1. *Introduction*

A satellite channel is particularly susceptible to errors. Errors occur frequently in satellite links, due to humidity (which causes attenuation and dispersion) and due to effects present in the wireless transmission channels (multi-path fading, shading, etc.). Although these effects are tolerable for frequencies below 10 GHz, they are substantial at higher frequencies. Moreover, the attenuation effect increases with frequency, thus transmissions in bands V and Q suffer more than those at band ka.

A way to tackle these errors is simply to transmit more power. This solution is not typically used for several reasons:

– designing a satellite for transmission, using more power, involves a heavier and larger satellite which will increase the satellite's cost;

– there are limits on the satellite's size and weight imposed by launching the rockets;

– a high-power ground station also involves greater size, as well as greater cost and operating costs.

An alternative to using high power is to use antennae with a higher gain. This does not necessarily mean using a larger antenna, rather a narrow beam antenna on the satellite can be used to closely focus the power beamed from the satellite, and can similarly improve the satellite's capacity to receive a signal from a small part of the Earth's surface. Consequently, making a satellite beam smaller effectively increases the power of a signal to be received either by the satellite or a user [SOS 06].

Apart from the techniques mentioned above, there are two other techniques for correcting error, the forward error correction (FEC) technique and the automatic repeat request (ARQ) technique. In this work, we use convolutional encoding, which is an FEC technique method.

Convolutive codes, introduced in 1955 by Elias [ELI 55], form an extremely flexible and effective class of error correcting codes. These are the codes most used in fixed or mobile telecommunications systems. Convolutive codes are applied on infinite sequences of information symbols and generate infinite sequences of coded symbols [PSA 99].

8.2. Structure of convolutive codes

8.2.1. *Convolutive performance code 1/n*

Consider the binary linear block code specified by the following generative matrix:

$$G = \begin{bmatrix} \mathbf{g}_0 & \mathbf{g}_1 & \mathbf{g}_2 & \cdots & \mathbf{g}_{L-1} & & & & & & \\ & \mathbf{g}_0 & \mathbf{g}_1 & \mathbf{g}_2 & \cdots & \mathbf{g}_{L-1} & & & & & \\ & & \mathbf{g}_0 & \mathbf{g}_1 & \mathbf{g}_2 & \cdots & \mathbf{g}_{L-1} & & & & \\ & & & \ddots & & & & \ddots & & & \\ & & & & \ddots & & & & \ddots & & \\ & & & & & \mathbf{g}_0 & \mathbf{g}_1 & \mathbf{g}_2 & \cdots & \mathbf{g}_{L-1} \end{bmatrix} \quad [8.1]$$

Each element $\mathrm{g_l}$ $0 \le 1 \le \mathrm{L} - 1$, of the matrix is a binary vector line with n components of the shape:

$$\mathbf{g}_l = \left[g_{l1}, g_{l2}, \cdots g_{ln} \right] \quad [8.2]$$

The zones left empty on the bottom left and the top right correspond to completely non-existent elements. Each line is obtained by shifting the previous of n columns to the right. This matrix, of size $B\times(nB+L-1)$ thus describes a linear block code in which the code words could be generated by the procedure on Figure 8.1.

Code words are generally very long, that is to say B>>L, and we can even consider this matrix as infinite. This linear code therefore no longer really possesses any dimension or length. The performance cannot therefore be defined as for block codes. However, by neglecting the code's closure (emptying registers), we note that there is a $1/n$ between the number of information bits and the number of bits of the code word. The performance for this code is therefore: $R=1/n$.

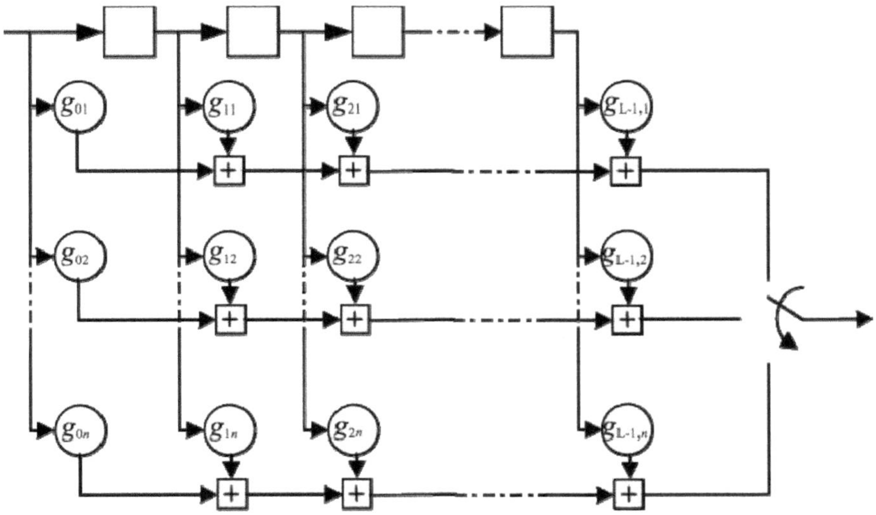

Figure 8.1. *Convolutive performance coder = 1/n*

The quantity L is called the convolutive code's *constraint length*.

The term 'convolutive' is applied to this class of codes because the coded bit sequence **s** can be expressed as the result of the information bit's convolution sequence **e** by the coefficients g. In fact, since the code is linear, we have: **s** = **e**. G. Observing the particular shape of G, the encoders' n output bits at the instant t, corresponding to an input, are written:

$$\mathbf{s}_t = \sum_{u=\max(1,t-L+1)}^{t} e_k \mathbf{g}_{u-k} \quad [8.3]$$

The encoder's complexity is independent of the length of the information bit sequence emitted and only depends on the code's constraint length.

8.2.2. *Convolutive performance code k/n*

To obtain a performance equal to a rational number any lower than 1, we must generalize the shape [8.1] of the generating matrix and the diagram of the encoder in Figure 8.1. We now suppose that k information bits are introduced in parallel at a given instant in the encoder or that a time offset of the vector e of information bits is made by a block of k bits. This comes to replace the line [g_0 g_1 … g_{L-1}] of size 1⊕L.*n* with a binary submatrix of size k⊕L.n. whose elements making up gl are of size k⊕n. The general construction by offsetting the matrix remains the same:

$$G=\begin{bmatrix} {}_k\updownarrow \overset{\overset{n}{\longleftrightarrow}}{\mathbf{g}_0} & \mathbf{g}_1 & \cdots & \mathbf{g}_{L-1} & & & & \\ & \mathbf{g}_0 & \mathbf{g}_1 & \cdots & \mathbf{g}_{L-1} & & & \\ & & \ddots & & & \ddots & & \\ & & & \mathbf{g}_0 & \mathbf{g}_1 & \cdots & \mathbf{g}_{L-1} \end{bmatrix} \qquad [8.4]$$

8.3. Polynomial representation

Take a convolutive performance code k/n with constraint length L. Its input can be seen as the data from k infinite bit sequences, which we represent using x series:

$$e_i(x)=\sum_{\ell=0}^{+\infty} e_{i\ell}x^{\ell}, \qquad i=1,\cdots,k \qquad [8.5]$$

where the unknown x symbolizes a delay of one period. The coefficient e_{il} is the bit present at the part th encoder's entry at instant l. In the same way, the encoder's n outputs can be indexed by j and are written:

$$s_j(x)=\sum_{\ell=0}^{+\infty} s_{j\ell}x^{\ell}, \qquad j=1,\cdots,n \qquad [8.6]$$

The relationship introduced between the inputs and outputs by the convolutive code can be written elegantly with its notations, by introducing generator polynomials. The generator polynomial that links the ith input to the jth output is written $g_{ij}(x)$, and is less than or equal to L-1. The r degree coefficient of this polynomial is worth 1 if there is a connection between output j and the bit present in the rth register corresponds to input i.

The octal is currently used to represent these polynomials. Thus, $g(x) = 1 + x + x^3$ will be represented by the number 15.

The outputs are calculated depending on the inputs with the help of the generator polynomial:

$$s_j(x) = \sum_{i=1}^{k} g_{ij}(x) e_i(x), \qquad j = 1, \cdots, n \qquad [8.7]$$

Since the time convolution corresponds to polynomial multiplication.

8.4. Graphic representations of convolutive codes

The idea of representing a convolutive code graphically comes from the Markovian characteristics of the output of the encode. In fact, the encoder's output depends on its input and its states. The graphs equivalent to the polynomial representation are often easier to manipulate and enable more powerful results to be obtained. Any convolutive code is represented by three equivalent but different graphs: the code's tree, the code's trellis and the state diagram.

8.4.1. *Tree diagram*

The coding can be represented by a tree diagram such that this corresponds to the encoder, each branch of the diagram corresponding to the coding of k information bits, and the n bits indicated on each branch representing the n bits of the code that are generated on the arrival of k bits of corresponding information. The diagram is traversed from left to right, and from there passes from one stage to the next by choosing a branch among the 2 k (L-1) branches. In these positions, suppose that the registers are initialized to zeros and we apply the k bits of the sequence to be coded to the encoder's entrance.

Studying a convolutional diagram using a tree diagram quickly becomes impractical as soon as the sequence to be coded exceeds several bits, as the tree's dimension doubles at each stage. It is therefore preferable to represent the encoder's performance with the help of a state diagram. Taking the example of a convolutional coding with a half-rate coding and constraint length 3, the corresponding tree diagram is shown below (Figure 8.2).

In this diagram, the tree is divided into two branches, the upper branch corresponds to a one bit input at 0 and the lower branch, to 1. The binary pair mentioned on each branch corresponds to the encoder's output. Once the tree's extremity is reached, we return to the beginning, using the two previous bits.

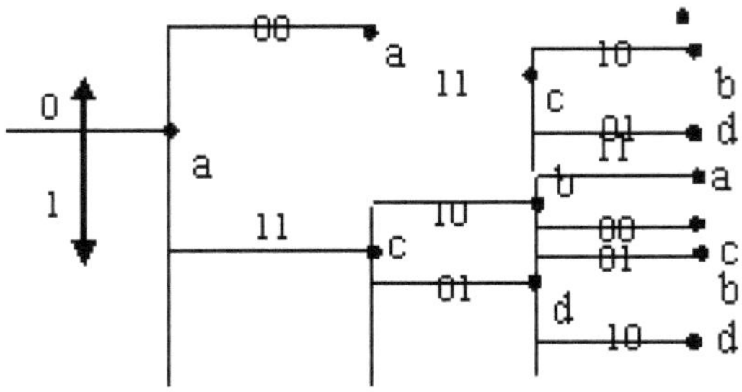

Figure 8.2. *Tree diagram for k = 1, n = 2 and L = 3*

8.4.2. *State diagram*

In this, each node represents a particular state of the encoder's register, and each branch represents the encoder's state change according to the arrival of new k bits. The different branches are identified by the value of the k bit input that triggers the state change and by the code word produced on the arrival of these k bits. In this usual case, there is a 2 k(L-1) state.

For the previous example, the corresponding state diagram is represented below.

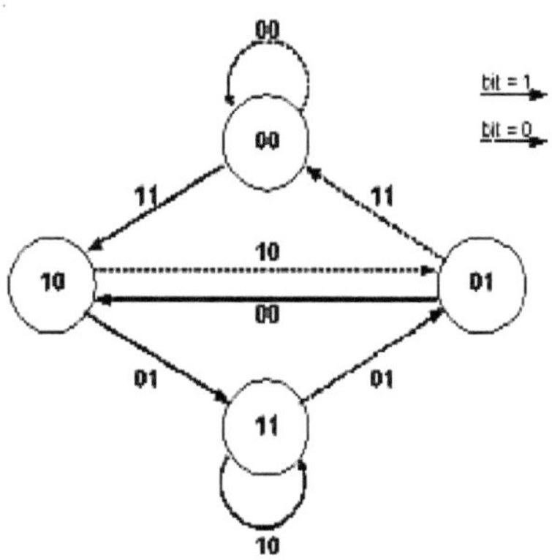

Figure 8.3. *State diagram for k = 1, n = 2 and L = 3*

The circles' content represents the register's state, the dotted arrows input bit 0 and the continuous arrows, input bit 1. The output's value is written on these arrows.

8.4.3. *Trellis diagram*

The path of the point on the graph (Figure 8.4) is made from left to right. Each transition from one state of the register to the following state is indicated by a line, and indicates the code word produced.

The automaton's internal state is resumed at stages (L-1), the input bits k produce the transition, whereas the content of the last stage is not significant as it will be eliminated by the shift. The (L-1)*k bits correspond to stage (L-1) defining possible state 2(L-1)*k , and at each node 2 k branches lead and 2 k branches depart. For the previous example, the corresponding state diagram is represented below:

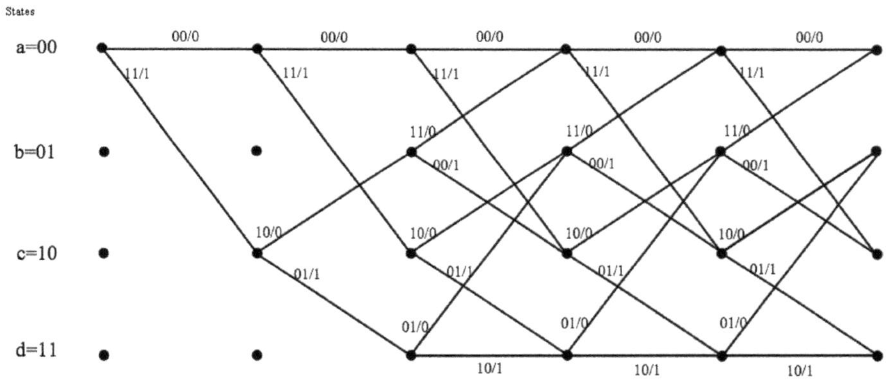

Figure 8.4. *Trellis diagram for k=1, n =2 and L= 3*

Regarding the state diagram, the continuous lines represent input bit 1, the dotted lines bit 0. The left-hand column corresponds to the register's state. From each step, we begin horizontally from the left-hand column.

The trellis enables the code's minimal distance to be found easily. It is a matter of the minimal Hamming weight between two code words, and so two paths. By comparing all the paths to the path at zero, in our case, a path diverging to any instant of the path at zero, by following *a→c→b→a* to converge again on the path at zero has the minimal output weight WHmin=5. We deduce from this that the minimal distance dHmin (also called the free distance) of the convolutive code (7,5) is equal to 5. This path corresponds to an input weight of 1.

8.5. Decoding algorithms

Decoding the channel (error correction coding) is a much more complex function. It represents one of the major problems in communications theory. There is no generic algorithm of acceptable complexity enabling any code to be optimally decoded. In the case of convolutional codes, there is the Viterbi algorithm. Its principal is the same as that of dynamic programming, and it is based on the notions of state and automaton. It uses the fact that the encoder corresponds to a very simple automaton, having a limited number of states. The receiver, to estimate the sequence emitted, will seek the path in a trellis that minimizes an additive distance criterion [VIT 71].

8.5.1. *The Viterbi algorithm*

The Viterbi algorithm is described in the context of maximum likelihood decoding (with firm or flexible input) convolutive codes. The description is similar for other applications of the VA, on the condition that the metric's definition is suitably modified.

Suppose that C is a binary convolutive code with performance R = k/n and constraint length L. The number of the encoder's input bits (and output bits respectively) is equal to k (respectively to n). The Viterbi algorithm is the most commonly used means of ML decoding when C has a reasonably-sized trellis, that is a number of states less than or equal to 256. The Viterbi algorithm is not complex and is simple to implant.

Maximum likelihood decoding corresponds to seeking the closest path (the most likely) and the sequence received (i.e. the observation) in code C's trellis. The distance used in the algorithm is either the Euclidian distance in the case of flexible input, or the Hamming distance in the case of fixed input [FOR 73].

Remember that the number of states in code C's trellis is equal to 2 k(L-1) and that the number of transitions by state is 2 k. Each summit in the trellis is connected on either side by 2 k branches towards the next and last vertices. The total number of branches between two stages in the trellis is 2 k(L-1) × 2 k = 2 kL. The sequence to be decoded being theoretically of infinite length, we are limited in practice to seeking the closest path on a window of width W. The paths manipulated in the code's trellis are therefore the length of W branches. If we try to apply a decoding method using an exhaustive search, and even if we suppose the initial and final states to be known in the case of a convolutive code whose trellis has been closed, we find ourselves confronted with seeking the best path among (2 kL)(W-L) possibilities. The complexity of such a proceeding is prohibitive since it is exponential depending on W. The Viterbi algorithm described below brings about a great reduction in the search's complexity, which becomes linear in W.

8.5.2. *Describing the Viterbi algorithm*

The time is written using the index t. A state in the trellis is written by s_t at the instant t, where

$$0 \leq s_t < S = 2^{k(L-1)} \qquad [8.8]$$

$T(s_{t-1},s_t)$ represents the branch at instant t in the trellis associated with the transition from state s_{t-1} to state s_t.

The sum of the square of the distances l associated with the branches between the trellis's path arriving at state s_t at instant t closest to the observation and any observation up to the same instant is called the aggregated metric summit s_t and is written ≤≤ (t, s_t).

Initialization: ≤≤(0,0)=0.0 and ≤≤(0,s)=+ for s $\overline{0}$. In fact, we know that the encoder starts from state 0.

At stage t, for each state s_t:

Branch $T(s_{t-1},s_t)$ is chosen such that the cumulative distance is minimal ≤≤(t, s_t) between the path selected and the observation until the time t:

$$\lambda\lambda(t, s_t) = \min_{s_{t-1}} \left[\lambda(t-1, s_{t-1}) + d^2\{T(s_{t-1}, s_t), \text{observation}\} \right] \qquad [8.9]$$

State s_{t-1}, which corresponds to this minimum, is called smin. The surviving path is then stored:

$$\text{Survivor}(t, s_t) = \text{survivor}(t-1, s_{min}) + T(s_{min}, s_t) \qquad [8.10]$$

Result of ML decoding (with a W delay): branch T $(t\text{-}W, s, s')$ is obtained, belonging to the survivor (t, s_t) with the smallest aggregate metric ≤≤ (t, s_t) among all the survivors.

The Viterbi algorithm therefore requires 2kL metrics to be calculated at each stage t, hence the W×2 kL complexity, linear in W. However, the complexity remains exponential in k and L, which limits use to small codes (kL from 7 to 10 maximum). The size W of the decoding window is taken in practice to around 5L. This value guarantees (empirically) that the survivors converge on a single path within the decoding window. The Viterbi algorithm therefore requires 2 k(L-1) aggregate metrics and 2 k(L-1) survivors of 5 kL bit length to be stored.

8.6. Discreet Wavelet Transform (DWT)

8.6.1. *Introduction*

Wavelet transform is a tool belonging to the signal processing domain. Its ancestor, the Fourier transform, has played a very significant role; it enables the signal's frequency composition to be explored and through its properties, allows filter operators to be applied to it. During this transformation, the signal is broken down on a set of base signals that are cosine, sine or imaginary exponential and form an orthogonal base (which enables redundant information to be deleted).

But very quickly, it became apparent that this composition was insufficient (especially for signal reconstruction and for identifying terms: the necessity of bi-orthogonality, which permits a better correlation between the signal components). In 1910, Haar introduced the first wavelet transform. The wavelet transform is a tool that cuts data, functions or operators into frequency components following a resolution adapted to scale [DON 95].

8.6.2. *Definition and property of the discreet wavelet transform*

Mathematically, in discrete wavelet transform, also called multi-resolution analysis, we seek to write all the functions f of L^2 as the limit of successive approximations, each being a smoothed version of f, with more and more concentrated smoothed functions.

DEFINITION 8.1.– *A sequence (Vj) of a closed subspace of L^2 (R) is a multi resolution analysis if six properties are verified:*

$$\forall (j,k)\in Z^2, f(t)\in V_j \Leftrightarrow f(t-2^j k)\in V_j \quad [8.11]$$

$$\forall j \in Z, V_{j+1} \subset V_j \quad [8.12]$$

$$\forall j\in Z, f(t)\in V_j \Leftrightarrow f(\frac{t}{2})\in V_{j+1} \quad [8.13]$$

$$\bigcup_{j=-\infty}^{\infty} V_j = L^2(R) \quad [8.14]$$

$$\bigcap_{j=-\infty}^{\infty} V_j = \{0\} \quad [8.15]$$

$\theta \in V_0$ exists such that $\{\theta(t-n)\}_{n\in Z}$ is a Riesz' basis of V_q. [8.16]

For a better understanding, we give these properties' intuitive interpretations.

The first [8.11] shows that *Vj* + 1 is the image of *Vj* by a factor 2 dilatation; in other words there is an underlying frequency grid in geometric progression. The second property [8.12] shows that for any *j*, *Vj* + 1 is a sub-space of *Vj*; e.g. a low resolution signal is also a high resolution signal.

[8.14] and [8.15] respectively show that *Vj's* intersection is reduced to 0 on L^2 (at minimal resolution, all the image is lost) and that the meeting of Vj is dense in L^2 (at infinite resolution, all the signals are reproduced perfectly). [8.13] ensures that the V_m spaces correspond to different resolutions whereas the translation invariance:

$$f \in V_m \rightarrow f(t-2^m n) \in V_m, \forall n \in Z$$

is a consequence of [8.1].

Using this definition, the basic properties of multi-resolution analysis can be shown below. To do this, we define the operator Pm of orthogonal projection on Vm.

8.6.3. *Orthogonality and biorthogonality: approximation space*

The approximation of *f* to resolution 2^{-j} is defined as the orthogonal projection P_{vf} on V_j. To calculate this projection it is necessary to find an orthogonal base for V_j. The following theorem gives support for orthogonalizing the Riesz' basis $\{\theta(t-n)\}_{n\in Z}$ and for constructing an orthogonal base for each *Vj* sub-space by translating and dilating a single function θ, called the scale function.

THEOREM 8.1.– If $\{V_j\}_{j\in Z}$ is a multi-resolution analysis and ϕ the scale function whose Fourier transform is worth [MAL 98, MEY 06]:

$$\hat{\varphi}(\omega) = \frac{\hat{\theta}(\omega)}{\left(\sum_{k=-\infty}^{\infty} \left|\theta(\omega+2k\pi)\right|^2\right)^{1/2}} \qquad [8.17]$$

if:

$$\varphi_{j,n}(t) = \frac{1}{\sqrt{2^j}} \varphi\left(\frac{t-n}{2^j}\right)$$

The $\{\phi_{j,n}\}_{n\in Z}$ family is an orthogonal base for V_j, regardless of $j\in Z$.

Using the result arising from this theorem, the orthogonal projection of f in Vj can be written by

$$P_{V_j} = \sum_{n=-\infty}^{\infty} \langle f, \varphi_{j,n} \rangle \varphi_{j,n} \qquad [8.18]$$

The scalar product

$$a_j[n] = \langle f, \varphi_{j,n} \rangle = \int_{-\infty}^{\infty} f(t) \frac{1}{\sqrt{2^j}} \varphi(\frac{t-2^j n}{2^j}) dt = f * \varphi_j^*(2^j n) \qquad [8.19]$$

The Fourier transform's energy ϕ is concentrated on the interval $[-\pi,\pi]$, as illustrated on Figure 7.2.5.1. for a cubic spline wavelet.

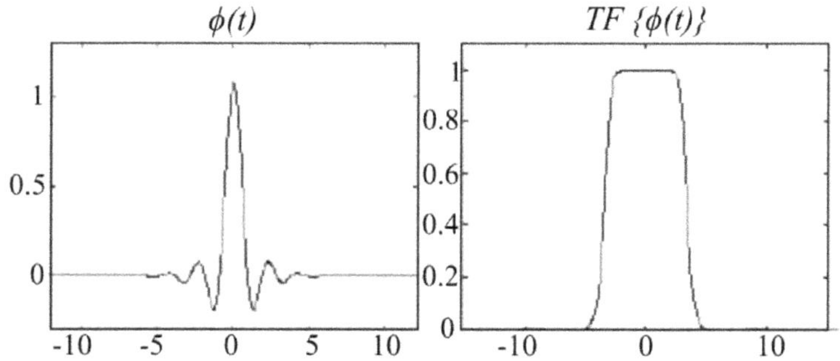

Figure 8.5. *Cubic spline wavelet function and its Fourier transform*

The Fourier transform $\sqrt{2^j}\hat{\varphi}^*(2^j\omega)$ of $\phi^*_j(t)$ is essentially centered on $[-2^{-j}\pi, 2^{-j}\pi]$. Consequently, the discreet approximation of $a_j\,[n]$ can be interpreted as the result of a low pass filter of signal f sampled at 2j intervals.

Nevertheless, the orthogonality imposes constraints that can be undesirable. One of the most significant is that an orthogonal scale function ϕ with compact support cannot be symmetrical and continuous. The property of symmetry is useful for analyzing finite-length signals. Some of these restrictions (notably on symmetry) can be attenuated by using biorthogonal multi-resolution analyses. In this way, a pair $[(V_j), (V^*_j)]$ of multi-resolution analyses forms a biorthogonal system if and only if

$$L^2(R) = V_0 \oplus (V_0^*)^{\perp} \qquad [8.20]$$

So, V^*_0 has a Riesz's basis with the shape θ^* (t-n), n describing Z, such that the translates of θ and θ^* form a biorthogonal system:

$$\left\langle \theta(t-k), \theta^*(t-k) \right\rangle = \delta_{n-k} \quad [8.21]$$

This is no longer described as an orthogonal base, but as a biorthogonal base.

8.7. Construction and discreet filtering

A multi-resolution analysis is entirely characterized by the scale function ϕ which generates an orthogonal base for all the Vj space. The property [8.12] implies that $\forall$ j $\in$Z, $V_{j+1} \subset V_j$. In particular $2^{-1/2}\phi$ (t/2) $\in V_L \subset V_0 1/2$. When $\{\phi\ (t-n)\}_{n\in Z}$ is an orthonormal base on V_0 we can write:

$$\frac{1}{\sqrt{2}}\varphi(\frac{1}{2}) = \sum_{n=-\infty}^{\infty} h[n]\varphi(t-n) \quad [8.22]$$

with

$$h[n] = \left\langle \frac{1}{\sqrt{2}}\phi(\frac{1}{2}), \phi(t-n) \right\rangle \quad [8.23]$$

The expression [8.22] is called the scale equation, and it shows that ϕ (t/2) is a linear combination of ϕ (t-n). The sequence h[n] can be interpreted as the impulse response to a discreet filter.

In the frequency domain, [8.22] is written according to:

$$\hat{\varphi}(2\omega) = \frac{1}{\sqrt{2}}\hat{h}(\omega)\hat{\varphi}(\omega) \quad [8.24]$$

with:

$$h(\omega) = \sum_{n=-\infty}^{\infty} h[n]e^{-j\omega}.$$

For p >0, the relationship [8.24] is written:

$$\hat{\varphi}(2^{-p+1}\omega) = \frac{1}{\sqrt{2}}\hat{h}(2^{-p}\omega)\hat{\varphi}(2^{-p}\omega) \quad [8.25]$$

By substitution, we obtain:

$$\hat{\varphi}(\omega)=\left(\prod_{p=1}^{P}\frac{\hat{h}(2^{-p}\omega)}{\sqrt{2}}\right)\hat{\varphi}(2^{-P}\omega) \qquad [8.26]$$

If $\hat{\varphi}(\omega)$ is continuous in $\omega=0$ then $\lim_{p\to+\infty}\hat{\varphi}(2^{-p}\omega)=\hat{\varphi}(0)$ and [8.26] becomes:

$$\hat{\varphi}(\omega)=\left(\prod_{p=1}^{P}\frac{\hat{h}(2^{-p}\omega)}{\sqrt{2}}\right)\hat{\varphi}(0) \qquad [8.27]$$

The study of h filters leading to multi-resolution approximations has given rise to many important theorems. The first of these gives the necessary and sufficient conditions for h to engender a scale function.

THEOREM 8.2.– [MAL 98], [MEY 06]

If $\phi\in L^2(R)$ is an integrable scale function. The Fourier series of $h[n]=<2^{-1/2}\phi(t/2), \phi(t-n)>$ satisfies

$$\forall\omega\in R, \left|\hat{h}(\omega)\right|^2+\left|\hat{h}(\omega+\pi)\right|^2=2 \qquad [8.28]$$

$$\hat{h}(0)=\sqrt{2} \qquad [8.29]$$

So, if $\hat{h}(\omega)$ verifies [8.28] and [8.29] and if the following conditions are verified: $\hat{h}(\omega)$ is periodic with period 2π and differentiable beside $\neq 0$;

$$\inf_{\omega\in[-\pi/2,\pi/2]}\left|\hat{h}(\omega)\right|>0 \qquad [8.30]$$

Then

$$\hat{\varphi}(\omega)=\left(\prod_{p=1}^{P}\frac{\hat{h}(2^{-p}\omega)}{\sqrt{2}}\right) \qquad [8.31]$$

is the Fourier transform of the scale function $\phi\in L^2(R)$.

This theorem represents a fundamental result in the context of the wavelet problem and above all for the practical implementation of wavelet techniques. In fact in practice, the scale function will be replaced by the coefficients h[n]. Moreover, as will be seen in the following section, this enables the wavelet transform and its derivations to be implemented using a filter bank structure. Condition [8.28] means that h is a conjugate mirror filter that will be defined in the following section. Condition [8.29] is a simple normalization whereas condition [8.30] enables it to be verified that the function defined by [8.31] has finite energy.

8.8. Defining the wavelet function: the place of detail

Orthogonal wavelets carry the details necessary for increasing the signal approximation. The approximations of f at scale 2 ^{and} 2^{j-1} are respectively equal to their projections on V_j and V_{j-1}. It is known that V_j is included in V_{j-1}. If Wj is the complementary part of V_j in V_{j-1}:

$$V_{j-1} = V_j \oplus W_j \quad [8.32]$$

The orthogonal projection of f on V_{j-1} can be broken down as the sum of the orthogonal projections on Vj and Wj:

$$P_{V_{j-1}} f = P_{V_j} f + P_{W_j} f \quad [8.33]$$

The projection $P_{Wj}f$ gives the details of f that appear at scale 2^{j-1} but which disappear at scale 2^j. The following theorem proves that an orthogonal base of Wj can be constructed by the translation and scale modification of a wavelet ψ.

THEOREM 8.3.– [MAL 98, MEY 06]

ϕ is a scale function and h – the conjugate mirror filter's impulse response. If ψ is a function whose Fourier is given by

$$\hat{\Psi}(\omega) = \frac{1}{\sqrt{2}} \hat{g}(\frac{\omega}{2}) \hat{\varphi}(\frac{\omega}{2}) \quad [8.34]$$

with

$$g(\omega) = e^{-i\omega} \hat{h}^*(\omega + \pi) \quad [8.35]$$

it is

$$\Psi_{j,n}(t) = \frac{1}{\sqrt{2^j}} \Psi(\frac{t - 2^j n}{2^j}) \quad [8.36]$$

for each of the scales 2^j, $\{\psi_{j,n}\}_{n\in Z}$ is an orthonormal base of Wj. For all the scales $\{\Psi_{j,n}\}_{(j,n)\hat{\imath}Z^2}$ is an orthonormal base of $L^2(R)$.

8.9. Wavelets and filter banks

In this section, using the concept of multi-resolution analysis, we suggest a cascade algorithm for calculating the coefficients resulting from the breakdown on an orthonormal wavelet base. In this way, we show a procedure for making the filters used in this algorithm and, following on, we determine the conditions that should be imposed on this structure so that the signal's reconstruction is perfect.

8.9.1. *Fast Wavelet Transform (FWT)*

The calculation for this transformation is based on a filter bank structure which one by one breaks down each approximation $P_{Vj}f$ into a smoother approximation $P_{Vj-1}f$ and the coefficients $P_{Wj+1}f$ corresponding to the information difference between two "versions" of f at two consecutive levels of resolution.

When $\{\Psi_{j,n}\}_{n\in Z}$ and $\{\varphi_{j,n}\}_{n\in Z}$ are orthonormal bases for Vj and Wj, the projections in these spaces are characterized by:

$$a_j[n] = \langle f, \varphi_{j,n} \rangle \text{ and } d_j[n] = \langle f, \Psi_{j,n} \rangle \qquad [8.37]$$

The following theorem shows that these coefficients are calculated by a cascade of discreet convolutions and down-sampling procedures. The result below shows great importance, as it provides us with a very effective practical tool, for calculating the TFR.

THEOREM 8.4.– [MAL 98, MEY 06]

The coefficients defined by [8.37] are determined by:

$$a_{j+1}[p] = \sum_{n=-\infty}^{\infty} h[n-2p]a_j[n] = a_j * \bar{h}[2p] \qquad [8.38]$$

$$d_{j+1}[p] = \sum_{n=-\infty}^{\infty} g[n-2p]a_j[n] = a_j * \bar{g}[2p] \qquad [8.39]$$

On reconstruction we have:

$$a_{j+1}[p] = \sum_{n=-\infty}^{\infty} h[n-2p]a_{j+1}[n] + \sum_{n=-\infty}^{\infty} g[p-2n]d_{j+1}[n] = \bar{a}_{j+1} * f[p] + \bar{d}_{j+1} * g[p] \quad [8.40]$$

The following notions have been used $\bar{x}[n] = x[-n]$ and

$$x[n] = \begin{cases} x[p], \, n = 2p, \\ 0, \, si \quad n = 2p+1 \end{cases} \quad [8.41]$$

where x symbolizes either h, or g.

On the following figure, the operations necessary to calculating coefficients a_n and d_n [MEY 06] are shown schematically.

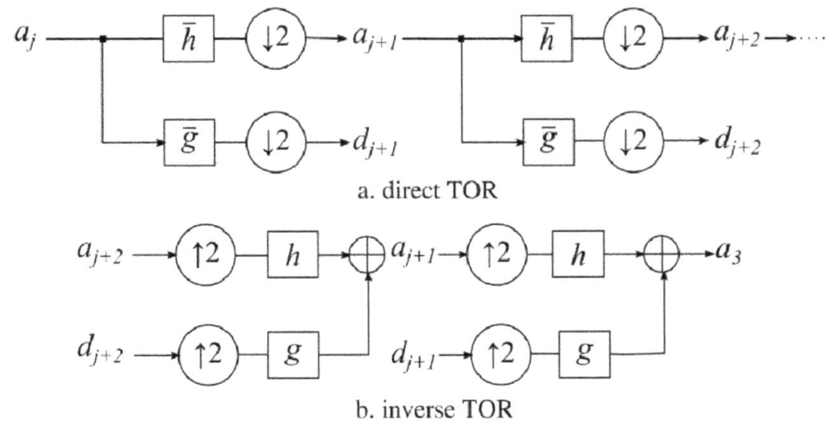

Figure 8.6. *Direct and inverse fast wavelet transform*

The symbol f represents the down-sampling operation: only one in two of the signal's points are taken. The orthogonal wavelet representation of aL=<f, $_{L,n}$> is composed of the detail coefficients of f at scales $2^L<2^j<2^J$ plus the approximation coefficients at scale 2^J:

$$\left[\{d_j\}_{L<j\leq J}, a_J\right] \quad [8.42]$$

THEOREM 8.4.– provides an optimal technique, from the point of view of implementation, for calculating the TOR. From an initial vector containing 2^N samples (in practice, for a minimal calculation time, it is preferable to have signals of a dyadic size), two sub-vectors of 2^{N-1} samples are created. One is obtained by low pass filtering (with the h filter) and the other by high pass filtering (with the g filter). This procedure is repeated on the low pass version. On the following figure the case of a vector of $N = 2^3$ samples is shown.

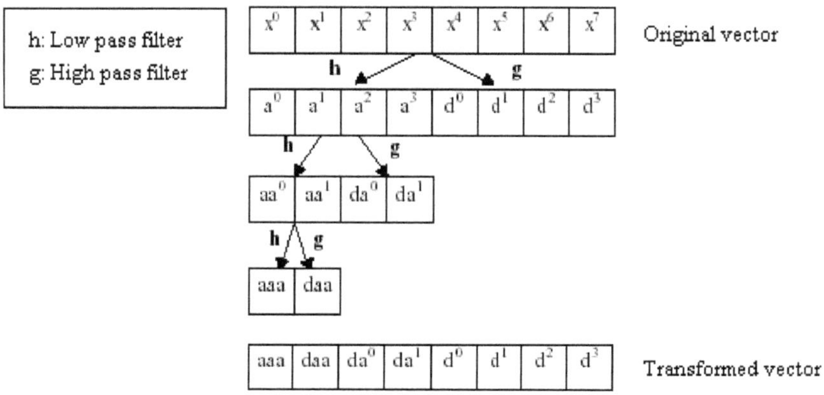

Figure 8.7. *The TOR for a vector of N=2^3 samples*

8.10. Thresholding coefficients

We will now tackle the different ways in which it is possible to modify the wavelet coefficients for noising purposes.

8.10.1. *General principle*

The wavelet coefficients correspond to a signal's details. When a detail is weak, it can be ignored without this visibly affecting the data. Thresholding wavelet coefficients are therefore a good means of eliminating the weakest details, which are considered to be noise.

In our case, a denoising is carried out, which consists only of conserving the most significant wavelet coefficients in absolute values, then reconstructing the signal from only these coefficients, the others having been set to zero. As the weakest details (i.e. the noise) have been eliminated, the signal appears "clean". We proceed in the following way [MAL 98].

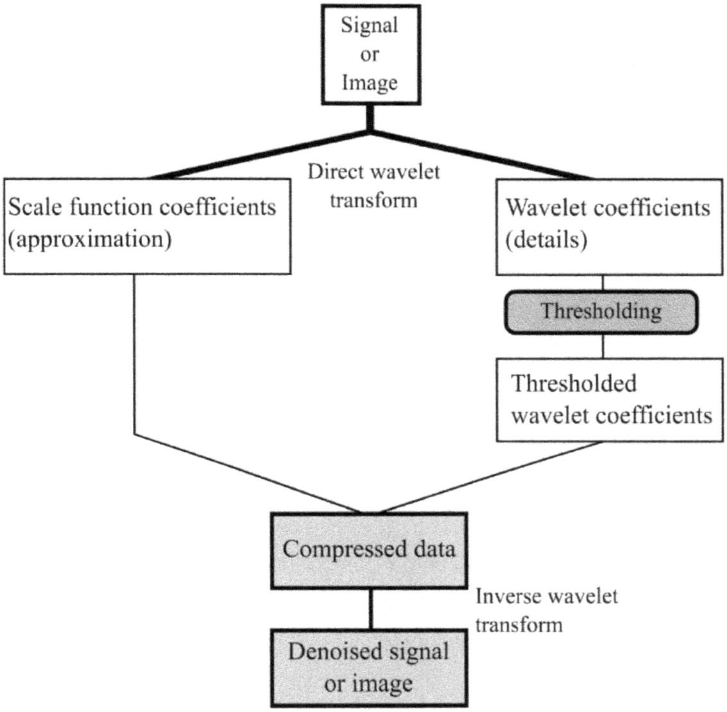

Figure 8.8. *Diagram of the threshold principle*

8.10.1.1. *Hard thresholding*

The principle of hard thresholding is simple. We start by fixing a threshold $\lambda > 0$. If the absolute value of a given wavelet coefficient d_j is higher than λ, we keep this coefficient as it is. If not, we set it to zero:

$$d_j \to \begin{cases} 0, \; si & |d_j| < \lambda \\ 1, \; si & |d_j| \geq \lambda \end{cases} \qquad [8.43]$$

We will see further how to determine the threshold λ .This transformation can be shown in a diagram as follows:

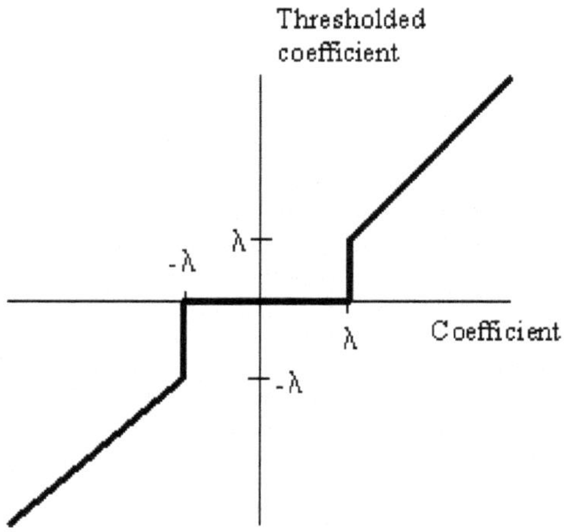

Figure 8.9. *Hard thresholding*

8.10.1.2. *Narrowing or "soft thresholding" or "shrinkage"*

Rather than simply "keeping or killing" a coefficient, narrowing applies the following transformation to each coefficient:

$$d_j \rightarrow \begin{Bmatrix} 0 & if|d_j| < \lambda \\ d_j - sgn(d_j) & if\ |d_j| \geq \lambda \end{Bmatrix} \quad [8.44]$$

which enables the coefficients to be thresholded more "softly":

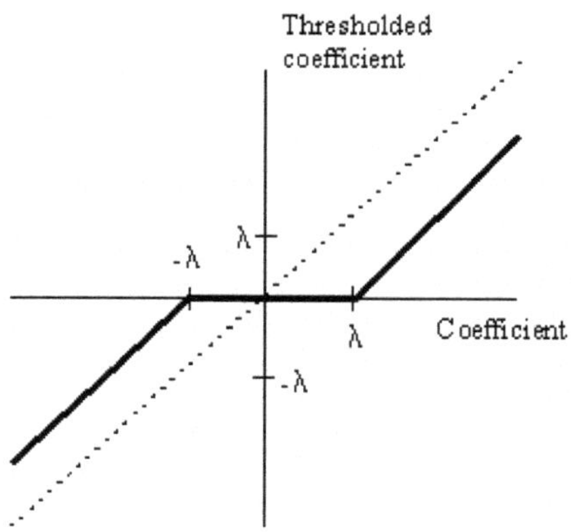

Figure 8.10. *Soft thresholding*

8.10.2. *Determining the threshold to use*

8.10.2.1. *Absolute threshold*

The value ≤ to be used is known (or found by trial and error). This is of course the simplest instance, but not the most common or realistic!

8.10.2.2. *Relative threshold*

The value of the threshold λ is defined as a fraction of the largest coefficient's absolute value. For example, all the coefficients are eliminated whose absolute value is less than $\sigma = 50\%$ of the largest coefficient's absolute value:

$$\lambda = \frac{\sigma}{100} \sup(\text{coefficients})$$

8.10.2.3. *Absolute quantitative threshold*

If we know the number N of wavelet coefficients resulting from the direct transform, we can directly choose the number N' of coefficients that we wish to keep. In this case, we keep the largest N' coefficients, and set all the others to zero. This selection is made by first classing the coefficients in decreasing order of absolute value, then by selecting the largest N'. The value of the last coefficient gives the threshold to be used.

8.10.2.4. *Relative quantitative threshold*

This is the most intuitive method: we choose to keep only σ % of the total number of coefficients (the largest in absolute value). The value of the threshold λ_1 is therefore determined by classing the coefficients in decreasing order, so that only the desired percentage remains.

If N is the number of coefficients to be thresholded and M the number of non–null coefficients after thresholding, we have: $\lambda = \dfrac{M}{N}$

This method is much used as the value of σ is directly linked to the compression rate. For example if $\sigma = 50$, only half the coefficients are kept, so the compression rate is close to 2 (by omitting the coefficients' positions). If $\sigma = 25$, a quarter of the coefficients are kept with a compression rate equal to 4.

8.10.2.5. *Universal or "Donoho threshold"*

Donoho and Johnston propose to threshold a set of N coefficients and to use the following value (by hard thresholding or narrowing) for the threshold λ:

$$\lambda = \sigma \frac{\sqrt{2 \log N}}{\sqrt{N}} \qquad [8.45]$$

Where σ here is a factor that represents the noise's variance. It can be shown that this method enables the difference between the reconstructed signal and the original signal to be minimized [DON 95].

8.10.3. *How do we separate and threshold the coefficients?*

The details are directly available, classed in increasing order of resolution on the lower right bound of the multi-resolution breakdown module. They depend on two variables: the position (t) and the breakdown level (i). The procedure to follow for the thresholding is therefore:

1) select details to form a single vector at a dimension t;

2) threshold the vector obtained by a given method;

3) range the thresholded coefficients again by level of breakdown I;

4) regroup the approximation and thresholded coefficients;

5) calculate the transform in inverse wavelets.

8.11. Simulating results

8.11.1. *Performance of convolutional encoding*

After a description of convolutional encoding, in this section, we show an analytical approach to convolutional encoding performances on chaotic DS-CDMA. Our analysis is based on using Gaussian approximation and bit error probability. We assume a perfect synchronization between transmitter and receiver.

When the errors are produced at the Viterbi decoder's output, this means that the trajectory has been badly chosen from distance d in the trellis, which occurs with a probability P_d. There are in total e_i erroneous trajectories, and there are d erroneous bits linked to this event. Adding through all the distance d, we obtain the erroneous bits' probability expression, expressed as follows:

$$P_e < \frac{1}{n} \sum_{d=d_f}^{\infty} d.a_d.P_d \qquad [8.46]$$

where d_f is the code's free distance. For a Gaussian channel, P_d is derived in [VIT 67]:

$$P_d = \frac{1}{2} erfc(\sqrt{R.d.SNIR}) \qquad [8.47]$$

where R is the encoding rate.

From [8.44] and [8.45] we have:

$$P_e = \frac{1}{2k} \sum_{d=d_f}^{\infty} d.a_d.ercf(\sqrt{R.d.SNIR}) \qquad [8.48]$$

If we suggest:

$$W_d = d.a_d \qquad [8.49]$$

P_e's expression becomes:

$$P_e = \frac{1}{2k} \sum_{d=d_f}^{\infty} W_d.ercf(\sqrt{R.d.SNIR}) \qquad [8.50]$$

For a large value of *SNIR*, the asymptotic performance is given by:

$$P_e = \frac{W_d}{2k}.ercf(\sqrt{R.d_{free}.SNIR}) \qquad [8.51]$$

8.11.1.1. *For a single user*

The *BER* vs *SNR* is simulated, for a variation of *SNR* from -25 dB to -5 dB with a step *of 2 dB* and with a spreading gain $N = 63$. The length of the constraints used for the coding is $L = 3$ and the coding rate $R = 1/2$. The first verification that can be seen in Figure 8.11 is that there is no large difference between the BER to be simulated and theoretical BER, which validates this theory.

The second verification is that there is a great improvement in performance, for a non-coded channel the BER = 10^{-5} for an SNR = -5.42 dB, whereas with convolutional coding the same BER is obtained for a SNR = -9.79 dB. But despite these performances, we see that this coding is incapable of correcting the errors triggered by a burst noise, which is, for an SNR = -25 dB, the BER = 0.38 whereas for a non-coded channel, the BER = 0.302.

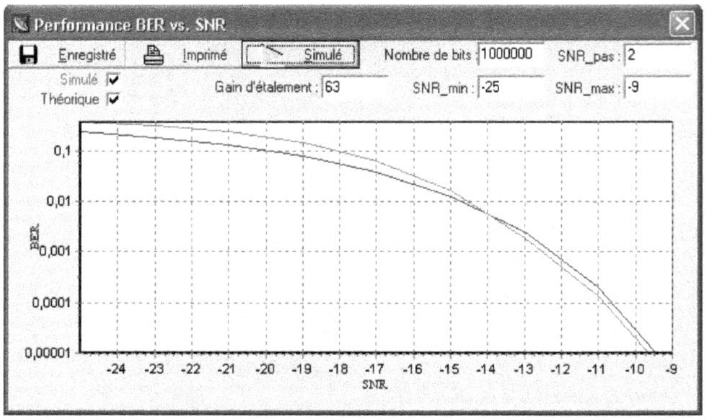

Figure 8.11. *BER according to the SNR with N=63 for a coded channel. For a color version of this figure, see www.iste.co.uk/benslama/quantum.zip*

8.11.1.2. *For multi-user*

After simulating a single user in the channel, the BER is now simulated according to the number of users for a SNR=0 dB, the users' variation step is fixed at 5 with a minimal value of 10 and a maximal value of 51.

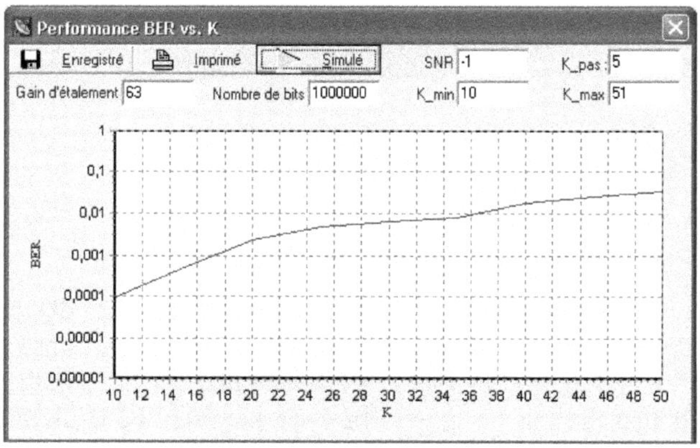

Figure 8.12. *BER according to K with SNR = 0 dB for a coded channel. For a color version of this figure, see www.iste.co.uk/benslama/quantum.zip*

As can be seen in Figure 8.12, each time that the number of users occupying the channel increases, the bit error probability increases (the transmission quality degrades), so there is a great improvement in performance compared to the non-coded channel, that is for a non-coded channel with K=10, the BER = 0.00455429, whereas with convolutional coding the BER = 0.00008207.

8.11.2. *Performance of convolutional coding with wavelets*

Here we will show performances of the combination of these two approaches, the coding used has the following property: an encoding rate R=1/2 and a constraint length L=3. For the wavelet, the Haar wavelet, which is defined by the function ψ, has been used:

$$\psi(x) = \begin{Bmatrix} 1 & 0 \leq x \leq \frac{1}{2} \\ -1 & -\frac{1}{2} \leq x < 1 \\ 0 & \text{elsewhere} \end{Bmatrix} \quad [8.52]$$

and by the scale function:

$$\Phi(x) = \begin{Bmatrix} 1 & 0 \leq x < 1 \\ 0 & \text{elsewhere} \end{Bmatrix} \quad [8.53]$$

The wavelet coefficients corresponding to the signal approximations on exiting the low pass filter are given as follows:

$$a_i = \frac{1}{2}\sum_{j=0}^{N-1} c_{2i-j+1} f_j, \quad i = 0,1,\ldots,\frac{N}{2}-1 \quad [8.54]$$

with:

$$c_0 = c_1 = 1 \quad et \quad c_2 = c_3 = \ldots\ldots = c_{N-1} = 0 \quad [8.55]$$

The wavelet coefficients corresponding to the signal's details on exiting the high pass filter are given as follows:

$$d_i = \frac{1}{2}\sum_{j=0}^{N-1} (-1)^j c_{j-2i} f_j, \quad i = 0,1,\ldots,\frac{N}{2}-1 \quad [8.56]$$

The inverse is carried out for the signal reconstruction:

$$f = \sum_{i=0}^{N/2-1} \left((c_{2i-j+1} a_i) + ((-1)^j c_{j-1-2i} d_i) \right), \quad j = 0,1,\ldots,N-1 \quad [8.57]$$

The breakdown level is 2, the thresholding method used is the soft method with a universal threshold value "Donoho threshold".

8.11.2.1. *For a single user*

The *BER* vs *SNR* is simulated, for a variation of *SNR* from -25 dB to -5 dB with a step of 2 dB and with a spreading gain of $N=63$. It can be seen in Figure 8.13 that there is a great improvement in performance compared to the coded channel, in Figure 8.14 a comparison has been made between the three simulations, the advantage of this approach can be seen clearly, whereas convolutional coding is powerless to correct the errors for a SNR=-25 dB, the performance has even degraded, as for the method proposed, it has overcome this problem, while still offering the best results.

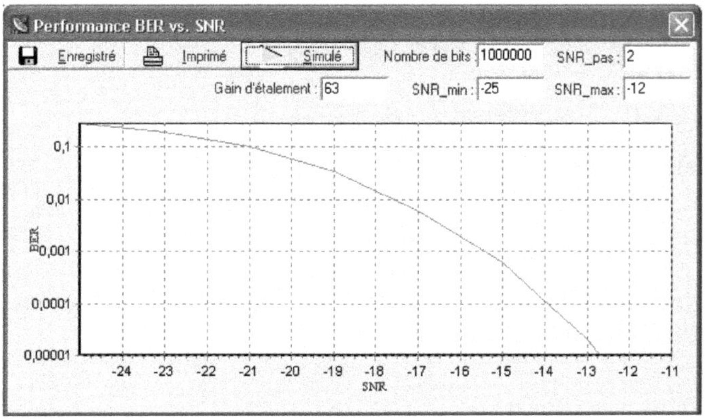

Figure 8.13. *BER according to the SNR with N=63 for the developed approach. For a color version of this figure, see www.iste.co.uk/benslama/quantum.zip*

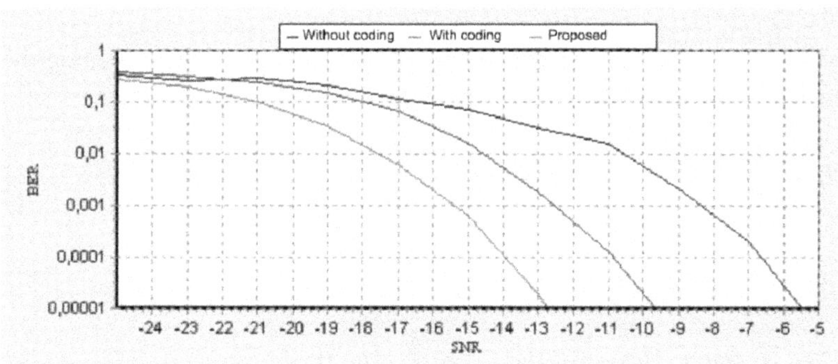

Figure 8.14. *Comparison of the BER according to the SNR for the three approaches. For a color version of this figure, see www.iste.co.uk/benslama/quantum.zip*

8.11.2.2. *For multi user*

After simulating a single user in the channel, the BER is now simulated according to the number of users for SNR = 0 dB; the users' variation step is fixed at 5 with a minimal value of 10 and a maximal value of 51.

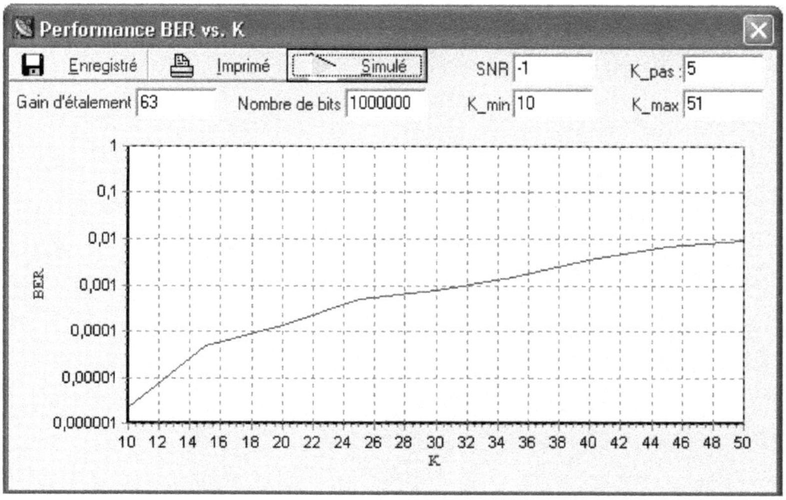

Figure 8.15. *BER according to K for SNR = 0 dB for the developed approach*

As can be seen in Figure 8.15, each time that the number of users occupying the channel increases, the bit error probability increases (the transmission quality breaks down), so there is a great improvement in performance compared to the coded channel. In Figure 8.16, a comparison has been made between three simulations; the proposed method's advantage and performance are obvious. This performance can be used to increase the number of users, while still ensuring a good quality of transmission.

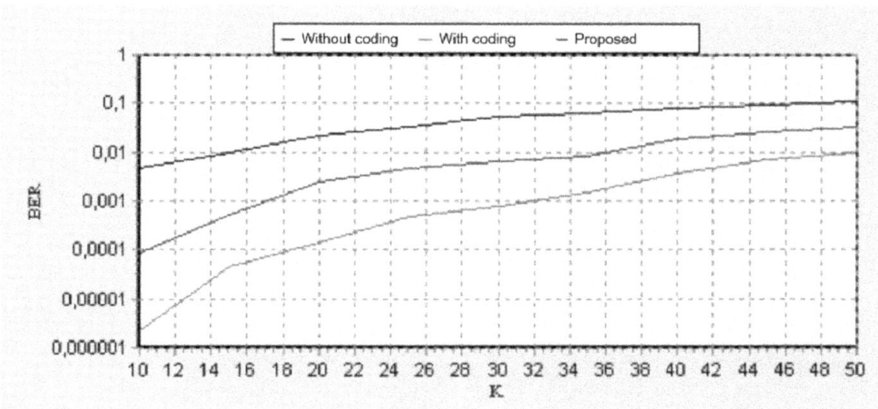

Figure 8.16. *Comparison of BER according to K for the three approaches. For a color version of this figure, see www.iste.co.uk/benslama/quantum.zip*

9

STAP Radar

9.1. Introduction

Detecting moving targets is one of the essential objectives of remote sensing. With a system for visualizing moving targets on the ground (or Ground Moving Target Indicator (GMTI)), an airborne radar receives the echo sent back by the target moving on the ground. However, the signal received not only contains the echo reflected by the target but also the echoes from the ground surface (illuminated or cluttered). These echoes come from all the areas illuminated by the radar's beam, so that they occupy all the distance intervals and all directions. The echo's total power is often much stronger than that of the desired signal, which poses a great challenge for detection. Filtering clutter is therefore an essential element in the GMTI system.

In traditional MTI systems, filtering clutter often takes advantage of the fact that the ground does not move. Consequently, the echoes are at zero Doppler in the Doppler spectrum. When the radar platform is also moving, the Doppler component is no longer zero. Moreover, Doppler components of the desired signal and that of the clutter are dependent. In this case, the clutter return is likely to have energy across the Doppler spectrum and the signal cannot be filtered. This is why a filter operating solely in the time domain can only effectively eliminate clutter at the price of a high attenuation of slow targets. These can, consequently, remain undetectable.

Scrambling is another source of significant interference that is often present in the received signal. The simplest form of scrambling is a jammer dam, which produces strong continuous white noise directed towards the radar receiver, which presents it from easily detecting the target. The jammer is generally linked to a specific direction. However, because of the nature of the white noise from the jammer, the scrambling signal received occupies the entirety of the Doppler band.

Space time adaptive processing (STAP) is a technique designed to solve these problems which has recently attracted the attention of numerous researchers in the radar community. In contrast to spatial or one dimensional time filtering, STAP simultaneously uses these two orthogonal dimensions in order to differentiate between the desired signals and the signals from interferences [HAI 96].

Figure 9.1 displays this filter, which can be seen from both spatial and temporal angles.

The axis labeled "temporal clutter filter" includes a filter formed from the inverse of the clutter spectrum projected onto the Doppler domain. Only the signals with a band stop Doppler frequency will pass through the filter. For example, a rapidly moving target will have a return signal with a high Doppler component and will not be canceled with the temporal filter.

In the case of spatial filtering, the inverse clutter spectrum is projected onto the azimuth domain. The spatial filter's band-stop which results from it is defined by the size of the radar's beam. The reflections from the ground clutter will be filtered by the spatial filter, which is advantageous for radars that aim to detect objects at a distance from the ground. However, all the targets of interest that are in the same direction as the ground clutter will also be deleted from the output signal.

9.1.1. *Space–time filtering*

STAP uses a combination of filters that operate in the spatial and temporal domains (azimuth and Doppler) simultaneously.

The notch from such a filter can be seen at the bottom left of the figure and coincides with the line of the clutter's Doppler–azimuth spectrum on the plot's diagonal.

It is clear from Figure 9.1 that no target will be blocked by the STAP filter, even if its Doppler and azimuth characteristics are very similar to those of the clutter.

Where there is jamming coming from a fixed source and a single direction, the spatial filter can be adapted to this signal by directing the radar's beam appropriately and attenuating the interference from the jammer. However, not all the targets of interest in the jammer's signal's direction will be detected. In a STAP scheme, using a filter enables targets that are from the same direction as the source of the jamming, but which have a different Doppler frequency to be detected.

STAP is therefore a high quality filtering system for a scenario where jamming is present.

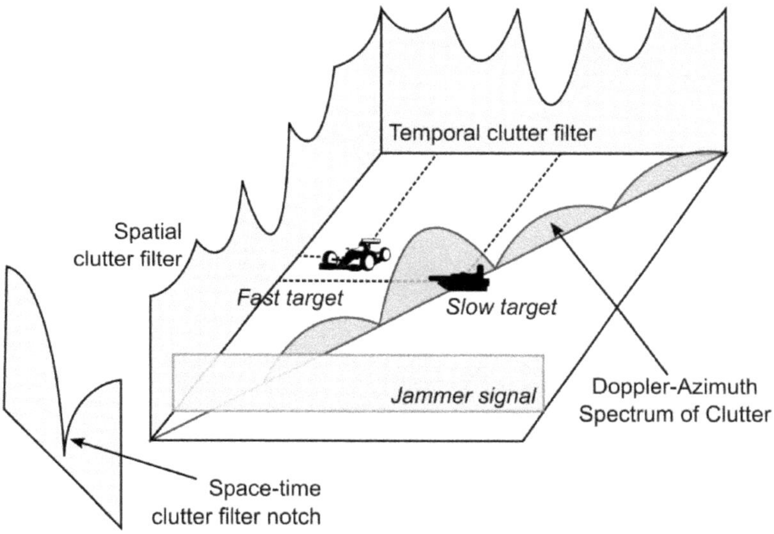

Figure 9.1. *Space–time filter*

9.1.2. *Geometric configuration*

The radar system being considered is a coherent, side looking pulse Doppler radar mounted on a mobile platform (airplane, satellite) at an altitude H and which moves at a uniform speed V_R as shown in Figure 9.2. The radar antenna is made up of N uniformly distributed elements separated by distance d and directed towards the platform's direction of movement. This radar emits bursts of M pulses, and L distance boxes are collected for each return period. The antenna is therefore side looking; this configuration is typical of an air–ground system on a high speed airplane. Here therefore, it is a question of detecting slow targets in competition with the echoes from the clutter coming from the antenna's main lobe and at speeds comparable to those of the targets, as illustrated in Figure 9.2 [BER 99].

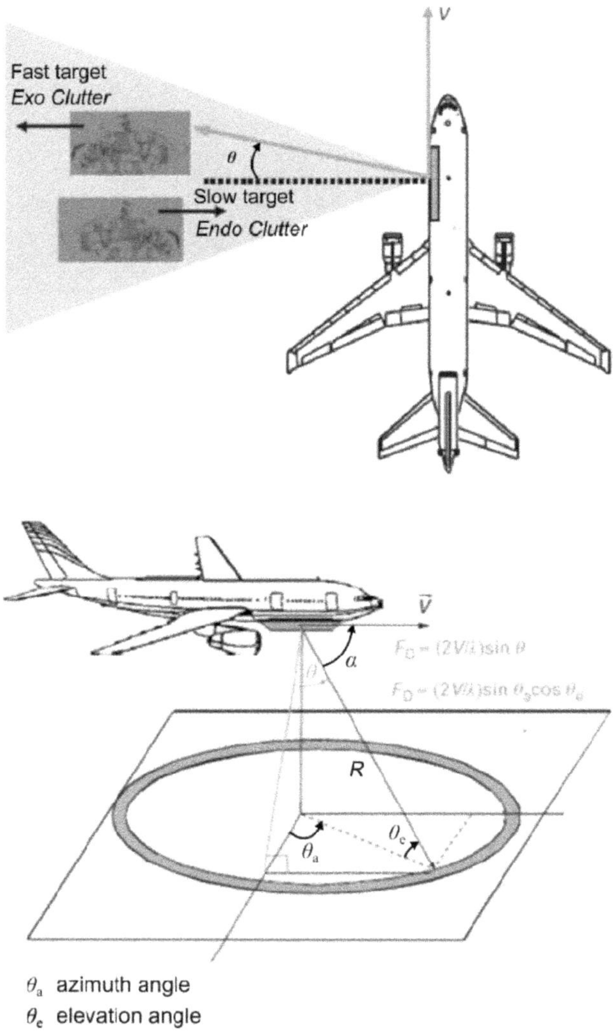

Figure 9.2. *Geometry of the configuration of an airborne side-looking radar*

9.2. Space–time adaptive processing (STAP)

STAP processing, which has keenly interested the radar community for a number of years, has the advantage of permitting a high rejection of clutter, and thus improves detection. Because of this it is being increasingly studied. Using optimized detectors in these processes should therefore permit further improvement in performance in terms of detection.

The basic principles of one-dimensional adaptive processing, as previously shown, can be extended to two dimensional processing (STAP); this process is defined as a linear combination that adds spatial samples of the elements of an antennae network and temporal samples of multiple pulses of a coherent wave form.

In order to calculate the STAP's adaptive weight vector, the environment statistics plus interferences are determined by forming a covariance matrix. Typically, this matrix is not initially known and should be estimated using the radar's secondary space–time samples [BER 99] (Figure 9.3). This estimation process supposes that the secondary data samples are independent and identically distributed (*i.i.d*). However, the real data recorded from the monostatic radar are not homogenous and non-stationary, thus violating the *i.i.d* data supposition.

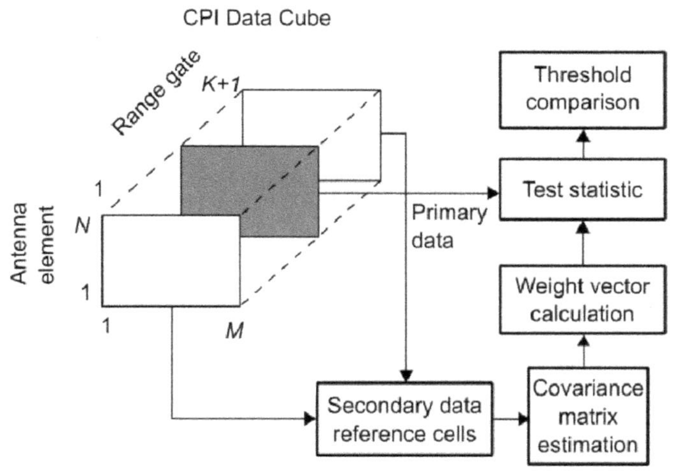

Figure 9.3. *Conventional STAP chain*

For each range cell, the adaptive weight is calculated from the resolution cells excluding X_N, the STAP's output, Y_N, will then be examined to test the presence of a target at an azimuth angle and a Doppler frequency.

The full-rank STAP methods use all the degrees of freedom available to eliminate interferences, thus requiring a higher calculation cost. In addition, this approach is based theoretically on the supposition that the noise is Gaussian i.i.d.

The airborne surveillance radar demands many degrees of freedom (DOF), which also leads to the need for a large number of samples that are also i.i.d.

In general, the number of samples required is less than the order of twice the number of the radar's degrees of freedom. Specifically, for a network of N elements and M pulses, 2MN identification samples are needed to achieve a performance corresponding to the SINR at a level of 3dB below the STAP optimum which cannot be available and feasible in real time [HAI 96]. The need for rank reduction methods, which provide an almost optimal interference suppression and a solution to the problem of high calculation cost, is of great interest. The concepts of rank reduction will be described in the next chapter.

9.2.1. *Mathematical data model*

Consider a space–time network with N uniformly spaced sensors and M delay elements for each sensor as indicated in Figure 9.4.

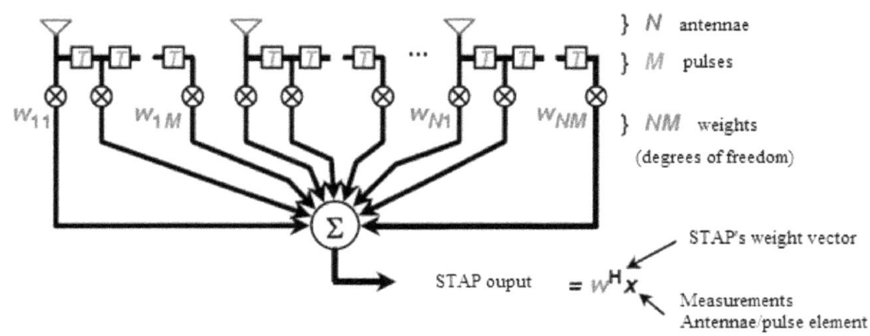

Figure 9.4. *General structure of the STAP filter. For a color version of this figure, see www.iste.co.uk/benslama/quantum.zip*

The radar transmits a coherent interval of M pulses to a constant *PRF*. For each *PRI*, the time samples (also called gate cells) are assembled to cover the gate interval, constituting a three dimensional model that is often called the STAP's "*data cube*", represented in Figure 9.5. For each gate cell, the radar observes the environment in two dimensions made up of N antennae elements and M pulses.

The data are therefore processed at a single gate of interest, which corresponds to a slice of the data cube. This slice is a space-time sample of NxM elements, each individual element corresponding to the data of the mth pulse repetition interval and the nth element of the antennae network.

In this space-time environment, the degrees of freedom provide domains corresponding to the "frequencies", the Doppler frequency and spatial frequency (azimuth angle) respectively [BER 99].

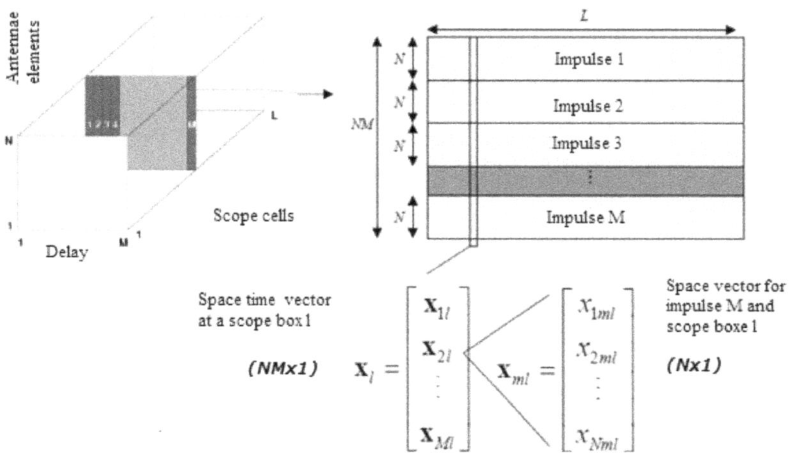

Figure 9.5. *STAP data cube*

The sample is stacked is the form of a column vector $\boldsymbol{X}$ given by:

$$X = \left[X_S^T(0), X_S^T(1), \ldots, X_S^T(M-1)\right]' \in C^{NMx1} \quad [9.1]$$

Where $\boldsymbol{X}_s(M) \in C^{Nx1}$ is the space sample containing all the spatial data corresponding to the mth pulse.

Under the hypothesis H_0, the data vector is the sum of the clutter's vectors $\boldsymbol{X}_c$, the jammers' $\boldsymbol{X}_j$ and the thermal noise $\boldsymbol{X}_n$. These vectors are assumed to be independent, complex and Gaussian with null expected value and they are not correlated with one another.

Under the hypothesis H_1, the component of the target's signal also appears in the data vector. By posing:

$$X_i = X_c + X_j + X_n \quad [9.2]$$

X_i represents the interferences that include the clutter, the jammer and the thermal noise. The two hypotheses are therefore respectively:

$$H_0 : X = X_i \quad [9.3]$$

$$H_1 = X_t + X_i \quad [9.4]$$

X_t is the target's signal and is given by:

$$X = \langle S \quad [9.5]$$

$\alpha = |\alpha| e^{j\phi}$ is a with random phase complex gain ϕ uniformly distributed between 0 and 2π, and $S \in^{NM \times 1}$ is the space–time direction vector, given by:

$$S = S_t \otimes S_s \quad [9.6]$$

where $\otimes$ represents the "Kronecker" product; $S_t \in C^{M \times 1}$ is the time direction vector and $S_t \in C^{N \times 1}$ is the spatial direction vector, given respectively by:

$$S_t = \left[1; e^{-j2\pi F_t}; e^{-j2\pi 2F_t}; \ldots\ldots\ldots; e^{-j2\pi(M-1)F_t}\right] \quad [9.7]$$

and

$$S_s = \left[1; e^{-j2\pi Fs}; e^{-j2\pi 2Fs}; \ldots\ldots\ldots; e^{-j2\pi(M-1)Fs}\right] \quad [9.8]$$

with $F_t = \dfrac{F_d}{PRF}$ and $F_s = \dfrac{d}{\lambda}.\sin\theta$ are the normalized Doppler frequency and the spatial frequency respectively, d is the space between the sensors, and θ is the azimuth angle.

9.3. Structure of the covariance matrix

The models for signals, noise and jammers mentioned above for SAP should be extended to STAP. The new space–time collector vector is used.

9.3.1. *Thermal noise*

The noise model in STAP is changed little from its origin in SAP:

$$R_n = E\left[X_n X_n^H\right] = \sigma_n^2 I_M \otimes I_N = \sigma_n^2 I_{MN} \quad [9.9]$$

Where I_{NM} is the identity matrix of dimension $N \times M$ and $\sigma\frac{2}{n}$ is the power of the thermal noise at the nth receiver. The thermal noise and the element are still modelled as a random Gaussian process, because these noises are spatially and temporally correlated.

9.3.2. *Jammers*

Unfortunately the jamming is spatially correlated, but not correlated from one pulse to another. Consequently, the covariance matrix of the jamming for a single jammer can be expressed in the form [PER 84]:

$$R_j = \sigma_{jn}^2 \left[I_M \otimes \left[S_{sj} S_{sj}^H \right] \right] \quad [9.10]$$

9.4. Clutter

For an airborne radar that moves with a speed V_R, the ground clutter received is considered to be a clutter ring with the same beam as the desired target signal, as shown in Figure 9.6. It is a complex signal because its energy is distributed in the angle–Doppler domain.

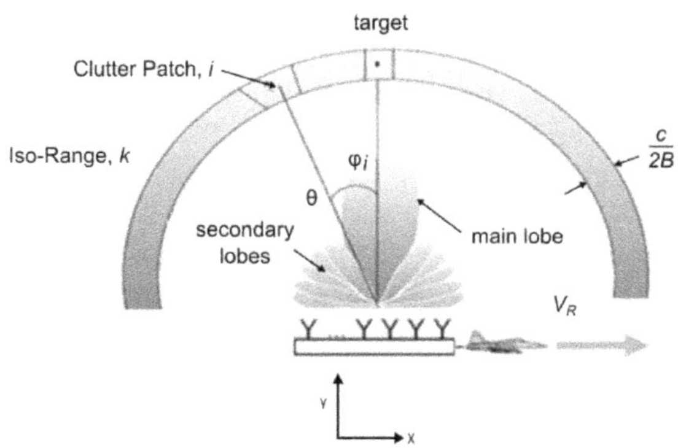

Figure 9.6. *Illustration of the clutter model*

It is assumed that the clutter can be approximated by dividing an angular region corresponding to a given gate into areas and calculating the echoes between these areas. The *PRF* is assumed to be low enough for there to be no ambiguity on the gate. For our simulations, it is assumed that there are 180 patches of clutter in this ring.

For a fixed gate, the area of each region of clutter can be described by its azimuth angle θ and the altitude angle Φ. Since the land is stationary, the clutter's relative land speed depends only on the platform's speed and its direction. The echo from each patch contains a Doppler shift due to the speed of the radar's platform, which can be calculated by the relationship:

$$f_c = \frac{2V_R}{\lambda}\cos\phi\sin\theta \tag{9.11}$$

The normalized Doppler frequency F_{tc} and the spatial frequency F_{sc} of the clutter are therefore expressed respectively by:

$$F_{tc} = \frac{2V_R}{PRF.\lambda}\cos\varphi\sin\theta \tag{9.12}$$

and:

$$F_{sc} = \frac{d}{.\lambda}\cos\varphi\sin\theta \tag{9.13}$$

$$F_{tc} = \frac{2V_R}{PRF \cdot d}F_{sc} = \beta F_{sc} \tag{9.14}$$

The clutter's location in the space-time domain is described by a straight line with slope β independent of the distance, as Figure 9.1 shows. This dictates the number of times that the clutter's Doppler spectrum "alias into" the non-ambiguous Doppler space [WIC 12]. Figure 9.7 corresponds to case $\beta = 1$ where there is no overlap. Figure 9.7 illustrates the shape of the clutter crest for different beta values.

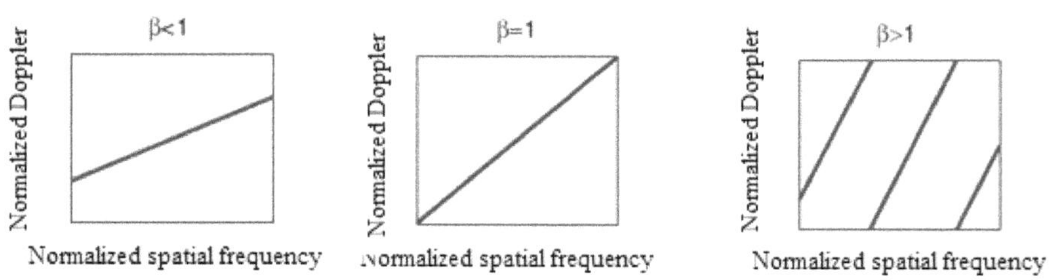

Figure 9.7. *Clutter crests for different values of PRF*

For β =1 and d/≤ =0.5, no ambiguity either over the Doppler or the gate.

For β =1 and d/≤ =1, unambiguous Doppler but ambiguous azimuth.

For β =0.5 and d/≤ =0.5, ambiguous Doppler but unambiguous azimuth.

The clutter is approximately a continuous field; the echo coming from each of the zones can be modeled as a superposition of a large number of discreet independent regions of the land N_c, equally distributed in azimuth angle around the radar.

$$X_c = \sum_{k=1}^{N_t} a_k \left(S_{tk} \otimes S_{sk} \right) \quad [9.15]$$

Where N_c is the number of clutter zones, α_k is the random amplitude of the kth clutter zone and S_{tk} and S_{sk} are respectively the temporal and spatial direction vectors that depend at the same time on the normalized Doppler frequency and the spatial frequency for each of the discreet land regions. Supposing that the echoes from different discreet land regions are not correlated, the clutter's space–time covariance matrix can be expressed by:

$$R_c = \sum_{k=1}^{N_t} \xi_k \left(S_{tk} S_{tk}^H \otimes S_{sk} S_{tk}^H \right) \quad [9.16]$$

where ξ_k is the clutter to noise ratio (CNR) by clutter zone. The total clutter to noise ratio would be:

$$CNR = \frac{\sum_{i=1}^{N_c} \sigma_c^2(i)}{\sigma_n^2} \quad [9.17]$$

In general, each clutter zone's power distribution depends on the azimuth angle, the spatial distribution and the clutter zone's radar cross section (RCS) [GUE 00].

These undesirable signals (clutter, noise and jammers) for STAP are assumed to be uncorrelated with one another. Consequently, the covariance matrix of interferences plus noise, R_i can be expressed as the sum of the individual covariance matrixes,

$$R_i = R_c + R_n + \sum_{i=1}^{N_J} R_J(i) \quad [9.18]$$

9.5. Optimal STAP

As described above, the STAP filter is an extension of the spatial formation in two space and time dimensions. The main objective is to maximize the filter's response to the useful signal while minimizing its response to interference (minimum variance distortion less response (MVDR)). In this section, the expression of the optimal STAP filter's weighting vector is given (in the sense in which the covariance matrix R_i is assumed to be known) as $\boldsymbol{w}_{opt}$ according to two optimization criteria, maximization of the SINR and the MVDR.

9.5.1. *Maximization of the signal to interference ratio*

Contrary to the SNR and JNR ratios, which describe the interferences at the element, the signal to interference to noise ratio (SINR) describes performances at the matrix, and more precisely the antennae network's exit. The SINR is similar to the SNR but takes into account the jammer effect.

One of the objectives of the detection theory is to maximize the probability of detection or to maximize the signal to noise ratio [GUE 00]. To this effect, the optimal filter $\boldsymbol{w}_{opt}$ can therefore be calculated by maximizing the signal to interference to noise ratio (SINR) as an objective function given by [RIC 01].

$$SINR_{LW} = \frac{\sigma_n^2 \left|W^H S\right|^2}{NMW^H R_i W} \qquad [9.19]$$

σ^2_t is the desired signal's power.

Assuming that $\boldsymbol{R_i}$ and $\boldsymbol{S}$ are known, the optimum weight $\boldsymbol{W}$ can be determined by maximizing relationship [9.18]. By using the Cauchy-Schwartz inequality, the optimum weight vector will be:

$$W_{opt} = \alpha R_i^{-1} S \qquad [9.20]$$

α is a complex arbitrary constant. The optimum *SINR* is obtained by substituting this vector in [9.18]. So:

$$SINR_{opt} = \sigma_t^2 S^H R_i^{-1} S \qquad [9.21]$$

9.5.2. *Minimizing the noise variance (MVDR)*

The optimum weight vector, given by [9.19], is obtained by supposing that the interference covariance matrix $\boldsymbol{R_i}$ is known. However in practice, the covariance matrix cannot be estimated using the data observed, as under hypothesis H_1, the target signal's component also appears in the data vector.

The MVDR criteria consist of reducing the output noise power to a minimum, subject to the constraint that the weight vector's response will have no distortion in the direction of the main lobe.

Where $\boldsymbol{R}_i = \mathrm{E}\{\boldsymbol{X}\boldsymbol{X}^H\}$ is the covariance matrix of the data observed.

By using Lagrange multipliers, the expression of the optimum weight can be obtained easily:

$$W_{MVDR} = \frac{R_i^{-1} S}{S^H R_i^{-1} S} \qquad [9.22]$$

If R_i is known, this weight vector is similar (almost to a factor) to that obtained by maximizing the *SINR* ratio. Therefore the weight vector *MVDR* is also optimal in the sense that the *SINR* is maximal.

9.6. Performance measures

9.6.1. *Beam pattern*

One of the performance measures commonly used to evaluate STAP algorithms is the STAP filter's angle–Doppler or simply beam pattern, defined as follows:

$$P = \left| W^H S \right|^2 \qquad [9.23]$$

For a radar with a uniform antenna and constant transmission frequency fr, the expression [9.23] is simply the inverse Fourier transform in two dimensions of the filter $\boldsymbol{w}$. Ideally, the filter's response should be at maximal gain in the angle–Doppler in the direction of the presumed target and zero in the direction of the interferences (clutter, jammers). Figure 9.8 illustrates the optimal filter's response (beam pattern) in the presence of two jammers at $-60°$ and $60°$ and a target at $\Delta t = 0°$ and a normalized Doppler frequency $\Delta t = 0.3$. There is a perfect cancellation of interference (clutter, jammers) and target extraction.

9.6.2. *SINR loss*

One performance measure for comparing the different STAP algorithms is SINR Loss: SINRLw. It is defined as being made up of losses caused by interferences at the filter's exit [ARM 02, RIC 01]. In other words, it is the relationship of the SINR at the STAP filter's exit w in the presence of interference plus noise at the SNR in the presence of thermal noise only. In this case, the filter is $W = \check{k}S$ where $\check{k}$ is a constant. The SINRL w is written as follows:

$$SINR_{LW} = \frac{\sigma_n^{\ 2} \left|W^H S\right|^2}{NMW^H R_i W} \qquad [9.24]$$

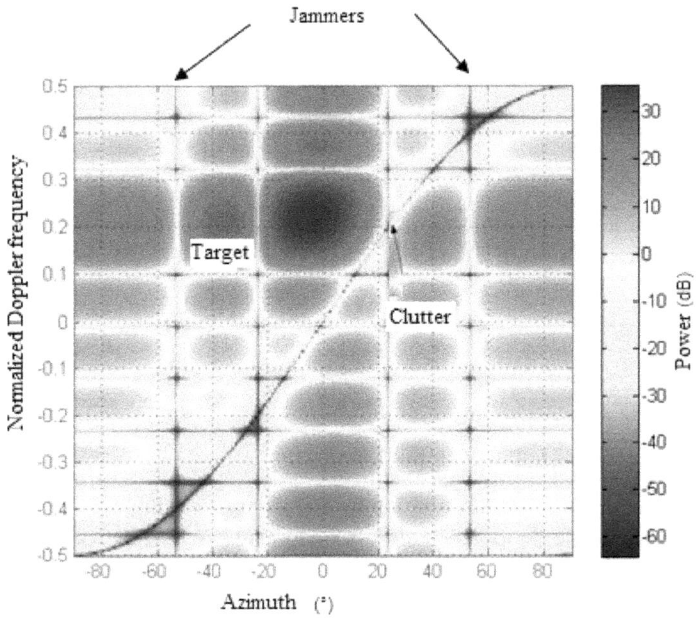

Figure 9.8. *The optimal STAP filter's response, at zero degrees of elevation, in the presence of two jammers at -60° and JNR=40 dB, N=12, M=10 and CNR=40 dB. For a color version of this figure, see www.iste.co.uk/benslama/quantum.zip*

9.6.3. *Improvement factor (IF)*

The performances of the algorithms shown in this work are discussed and evaluated by means of the improvement factor, *IF*. By definition, this factor represents the ratio between the *SINR* at the processors' exit and entrance. [BAD 05].

$$IF = \frac{W^H SS^H W_{tr}\left(R_i\right)}{W^H R_i WSS^H} \qquad [9.25]$$

For an optimal processor, the adaptive weight W is given by the expression [9.19]. The optimal improvement factor can therefore be obtained as follows:

$$IF_{opt} = \frac{S^H R^{-1} S S^H R^{-1} tr(R)}{S^H R^{-1} R R^{-1} \quad S S^H S} = S^H R^{-1} S \frac{tr(R)}{S^H S} \quad [9.26]$$

The optimal improvement factor is limited by [BAD 05].

$$\frac{1}{\lambda_{\max}} tr(R) \leq IF_{opt} \geq \frac{1}{\lambda_{\min}} tr(R) \quad [9.27]$$

where λ_{max} and λ_{min} are the proper maximal and minimal values of R respectively.

The clutter's notch (inverted peak) appears at the clutter's frequency in the direction of the radar. The size of this notch is a measurement of the detection of slow targets. It is mainly determined by the size of the network N, the length of the coherent processing interval M, and other effects of different noises.

9.7. Influence of the radar's parameters on detection

In this section, we will comment on the influence of some of the radar's parameters on detecting targets. These targets are supposed to be of low power ($SNR = 0$) and low speed (Table 9.1). All the environment and radar system parameters used for the simulations are summarized in this table.

The simulated radar is a side-looking linear network (parallel to the aircraft's longitudinal reference axis) formed of N antennae and M impulses in a coherent processing interval.

Symbol	*Description*	*Value*
ftr	Transmission frequency	0.3 GHz
c	Light speed	3.10^8 m/s
λ=c/ftr	Wave length	1m
V_R	Airborne radar speed	100m/s
d=λ/2	Space between elements	0.5m
N	Number of antennae elements	variable
M	Number of pulses	variable
σ^2	Variance of the radar's noise	1 (normalized)

Table 9.1. *Radar system parameters*

The elevation angle is fixed at 20°, and the azimuth angle represents the only free parameter.

The interference environment is made up of J jammers and land clutter. The azimuth angle's jammer to noise ratios (JNRs) as well as the clutter to noise ratios (CNRs) are variable. 180 patches in the clutter ring are considered.

9.7.1. *Influence of the number of antennae and impulses*

To show the effect of the number of antennae N and the number of impulses M on the detection of slow moving targets, it is assumed that the antennae are spaced at half the wavelength transmitted $d = \frac{\lambda}{2}$. Two examples have been considered: the influence of M with $N = 8$ and the influence of N with $M = 10$ with $RF = \frac{4V_R}{\lambda}$.

From Figure 9.9, we note that the increase in the number of impulses M leads to a decrease in the notch's size, until a fixed size is obtained. When M is greater than 10, the size of the notch remains constant despite the increase in the number of impulses. However, this increase demands a fairly considerable calculation time, and an obstruction in the material complicates the task in practice.

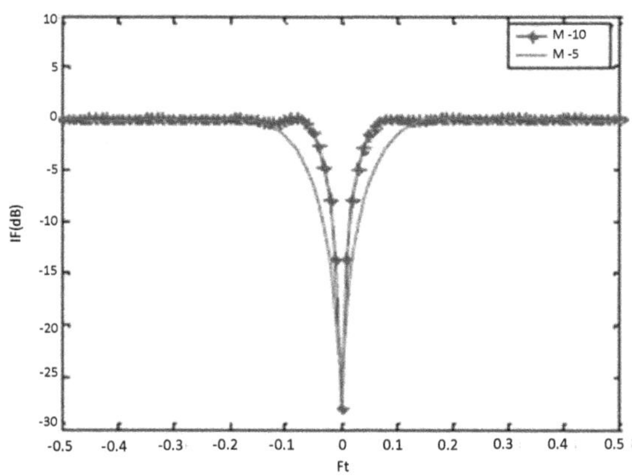

Figure 9.9. *Improvement factor for the optimal processor DFP. For a color version of this figure, see www.iste.co.uk/benslama/quantum.zip*

In the second instance, we note that the more the number of antennae increases, the straighter the notch (Figure 9.10.) But, from a number of antennae elements $N = 8$, the notch's length remains constant. Consequently, the same detection performance is obtained despite the obstruction and the complexity of the calculations, caused by the increase in the number of antennae elements N. It can be concluded that the improvement in detection can be accomplished by increasing a single parameter, the number of antennae or impulses or therefore the N/M ratio.

with $PRF = \frac{4V_R}{\lambda}, \frac{d}{\lambda} = 0.5$ and $N = 8$: M=5,10.

$$PRF = \frac{4V_R}{\lambda}, \frac{d}{\lambda} = 0.5 \text{ and } M = 10: N = 4, 10.$$

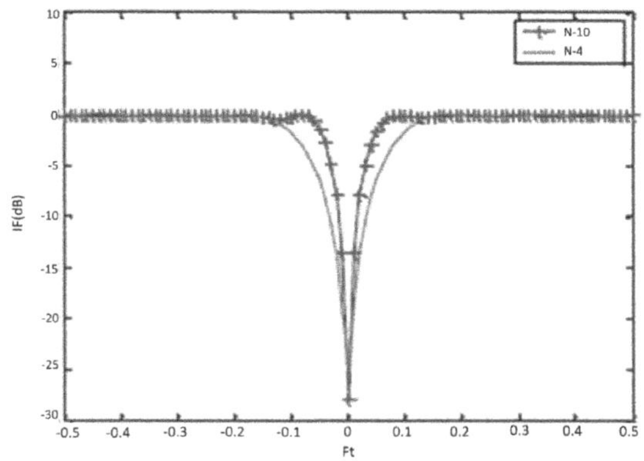

Figure 9.10. *Improvement factor for the optimal processor DFP with. For a color version of this figure, see www.iste.co.uk/benslama/quantum.zip*

9.7.2. *Influence of the spacing, d, between the antennae elements*

Figure 9.11(a) represents the improvement factor according to the normalized Doppler frequency for a scenario in which the antennae are spaced at a distance ($\frac{d}{\lambda}$ = 1), that is to say, double the Nyquist spatial sampling frequency. It is quite clear that ambiguous notches have appeared in the lateral bands; which is not the case for Figure 9.11(b), where the distance between the antennae is ($\frac{d}{\lambda}$ =0.5); that is to say equal to the Nyquist spatial sampling frequency. On this figure we can see the disappearance of the notches, which improves detection. Consequently, it can be said that the spatial sub sampling ($\frac{d}{\lambda}$ >= 0.5) leads to a spatial ambiguity or gate ambiguity, the detection of slow moving targets breaks down and becomes almost impossible.

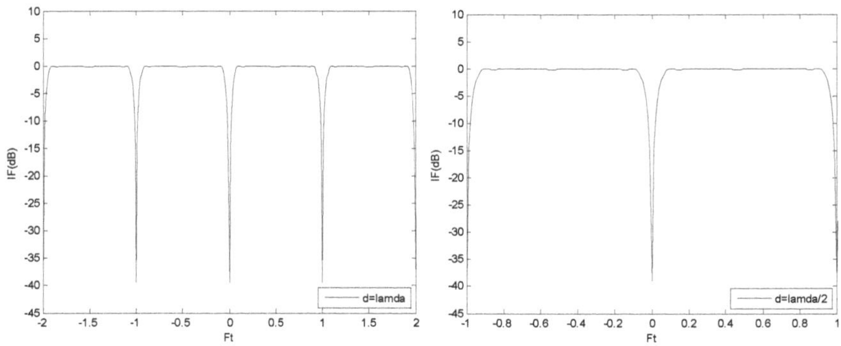

a) b)

Figure 9.11. *Improvement factor for the optimal processor DFP for different spacings: a)* $\frac{d}{\lambda} = 1$, *b)* $\frac{d}{\lambda} = 0.5$

9.7.3. *Influence of the PRF: temporal sub sampling*

From Figure 9.12, we can see that the notches are straight, which allows for the detection of slow moving targets. Also, there are ambiguous notches of clutter in which the associated velocities are called blind velocities.

In the case where the PRF is equal to $PRF = \frac{2V_R}{\lambda}$, we note the appearance of ambiguous notches and therefore the difficulty or indeed the impossibility of detection. But in the case of a PRF equal to $PRF = \frac{4V_R}{\lambda}$, we note that the notches are eliminated; there is an improvement in detection.

We can conclude that temporal subsampling leads to a temporal ambiguity (Doppler ambiguity), or even an overlapping of the PRFs.

This is due to the overlapping of the lateral bands, which becomes more significant with the decrease in PRF. Consequently, the effect of the Doppler ambiguities will be more significant and will make it more difficult to detect targets. The explanation will be more convincing if we consider the problem of sampling and the Nyquist frequency (Shannon's theorem). In fact, if we compare the two results obtained in (b) and (c) where we assumed a PRF to be equal to the Nyquist frequency: $PRF = \frac{4V_R}{\lambda}$ (*sampling*) and a PRF equal to half the Nyquist frequency: $PRF = \frac{2V_R}{\lambda}$ (*subsampling*) respectively, the well-known overlap phenomenon can clearly be seen manifested here in the appearance of echoes in the clutter's secondary lobes. This problem has been addressed by Klemm [KLE98] from another angle and has led to the same conclusion.

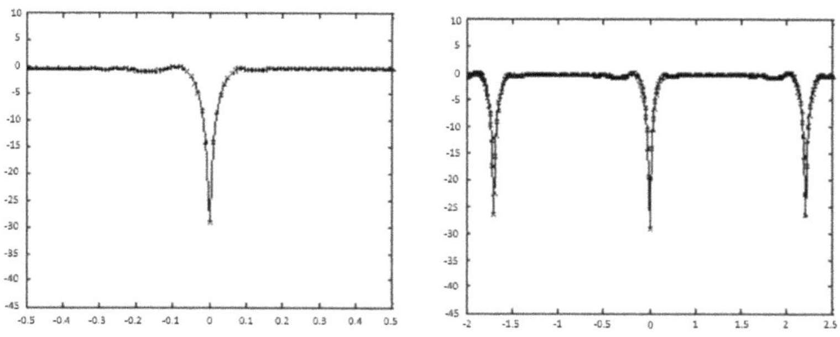

a) b)

Figure 9.12. *Improvement factor for the optimal processor DFP with PRF constant, N = 8, M = 10, d/λ = 0.5: a)* $PRF = \dfrac{4V_R}{\lambda}$; *b)* $PRF = \dfrac{2V_R}{\lambda}$

This analysis can be approached by varying the Beta factor, called the folding factor, defined as being inversely proportional to the PRF repetition frequency.

Figure 9.13 shows the received signal's minimum variance spectrum in the presence of two jammers for different values of β. We can see the presence of folding in the clutter's spectrum for high beta values (weak PRF). We can therefore say that the choice of PRF is paramount for ensuring better detection. It can be noted that each time that the PRF decreases, the clutter acquires other ambiguous notches.

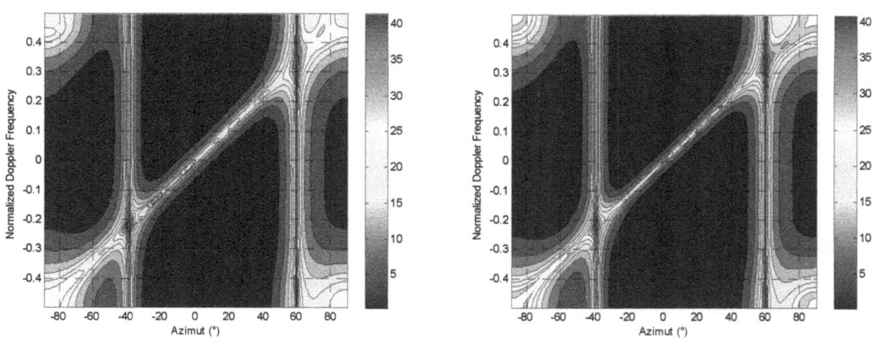

Figure 9.13. *Angle/Doppler spectrum in the presence of two jammers at -40° and 60° with JNR = 45 dB, N=8, M=10, CNR = 20 dB, β = 0.5, β = 1, β = 2. For a color version of this figure, see www.iste.co.uk/benslama/quantum.zip*

9.7.4. *Influence of the spatial frequency Bs: (Bs = 2d/λ)*

From Figure 9.14, we note that the more the clutter's spatial band increases, the more the ambiguous notches appear, which leads to a degradation in the detection of slow moving targets. This result is in agreement with Klemm's study [KLE 98]. In fact, he demonstrated that the size of the clutter's band is proportional to the number of the clutter's proper values and that these represent a measurement of the deletion filter's complexity.

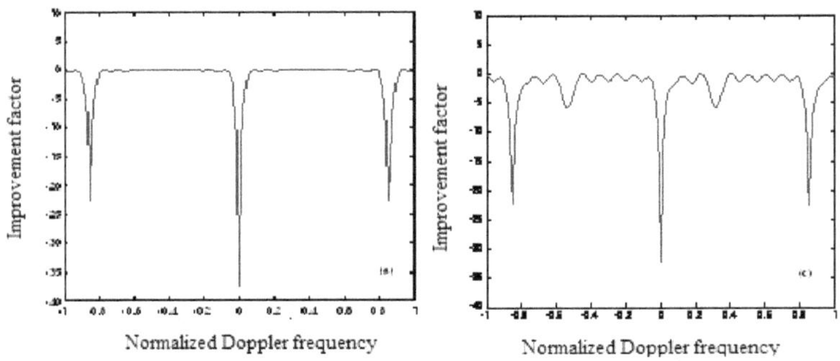

a) b)

Figure 9.14. *Improvement factor for the optimal processor DFP with* $PRF = \frac{4V_R}{\lambda}$; *a) Bs = 0; b) Bs = 0.1*

9.8. Sample matrix inversion algorithm (SMI)

The optimum weight vector in [9.22] assumes that the covariance matrix is known. In practice, the covariance matrix R_i is unknown and should be estimated using the test data. A standard estimation of R_i using a set of observations vectors $\{x_t\}$, $t = 1 \ldots L$ is given by:

$$\widehat{R_k} = \frac{1}{L} \sum_{t=1 t=k}^{L} x_t x_t^H \qquad [9.28]$$

The distance cell tested *k*, (primary cell) is left out of the estimate to avoid what is called self-nulling the target. Generally, the two cells neighboring the tested cell, called guard cells, are also excluded. Normally, the greater the number of formation cells, the better the estimation of the interference.

Under the hypothesis that the observation vectors (snapshots) are complex independent Gaussians and identically distributed (i.i.d), this estimator corresponds to the maximum likelihood estimator of the true covariance matrix R_k and it is distributed according to a complex Wishart law of degrees of freedom K and the matrix Rk. The algorithm based on this estimator is called sample matrix inversion (SMI) and the weighting vector is given by:

$$W_{smi} = \hat{k} R_k^{-1} S \qquad [9.29]$$

where ˜κ is a constant

In Figure 9.15, the normalized SINR's empirical distribution and the analytical distribution are traced. We see a perfect agreement between the two. This figure also shows the SMI's convergence according to the support, K. The SMI algorithm shows a slow convergence, for example for a SINR of -3dB below the optimal, K3 dB = 2 NM – 3 ~= 2 NM are necessary.

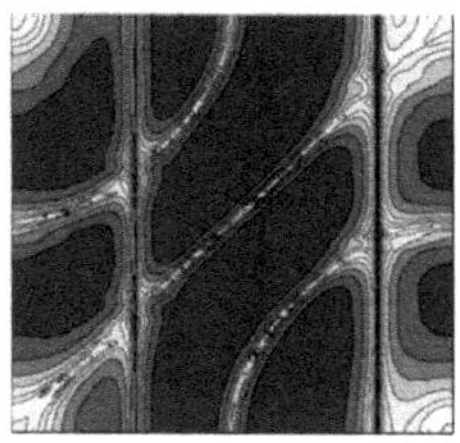

Figure 9.15. *This figure illustrates the filter's angle/Doppler response using the SMI algorithm. We note that the interferences are cancelled (clutter, jammer) but with a distortion of the secondary lobes by comparison to the optimal STAP. For a color version of this figure, see www.iste.co.uk/benslama/quantum.zip*

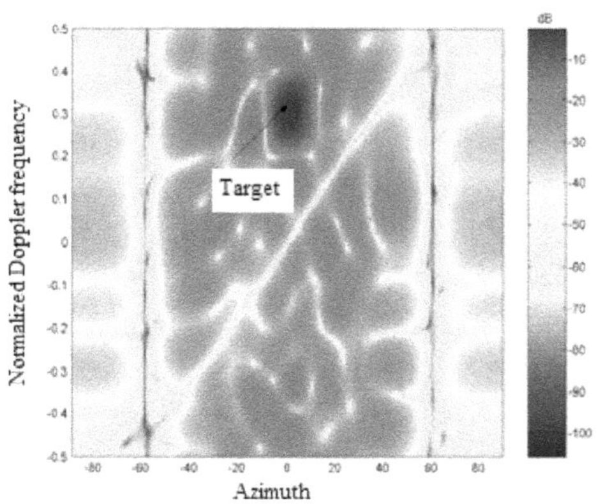

Figure 9.16. *The filter's angle/Doppler response using the SMI algorithm target inserted in cell 50, Ft=0.3 and in the presence of two jammers at JNR=40 dB and CNR=20 dB. For a color version of this figure, see www.iste.co.uk/benslama/quantum.zip*

9.9. Conclusion

In this chapter, we have illustrated the basic notions of adaptive one dimensional and two dimensional signal processing, while also mentioning the performance criteria to be aware of: the signal–interference–noise–ratio (*SINR*) and the improvement factor (*IF*) assuming a direct form processor. Finally, the results of the simulation are discussed while trying to show different radar parameters' influence on slow target detection performances. It has been shown that to achieve better detection in the environment considered, it is better to consider a number of impulses $M = 10$, and a number of antennae $N = 8$ spaced along half the transmitted wavelength. Quite clearly, in order to avoid the overlap phenomenon, choose a PRF greater than or equal to the Nyquist frequency. So take an PRF mean (beta=1) to avoid Doppler ambiguities and take (d= /2) to avoid gate ambiguities. Results on interference–noise subspace separation will be used in the algorithms for reducing the covariance matrix's rank. This will be detailed in the following chapter.

10

Tracking Radar (Using the Dempster–Shafer Theory)

10.1. Introduction

Multi-target tracking is a very important domain for aviation safety that requires great precision and continuity, so the quality of the tracking lies in good data combining. To achieve this in our work, we have elaborated on a theory [NIM 95] known for its effectiveness in combining and modeling imprecise data, called the Dempster–Shafer theory.

This method was used during the '80s for modeling uncertainty in expert systems and during the '90s in information fusion (remote sensing, target identification, medical imaging, etc.). This theory is a generalization of the Bayesian theory. It was introduced by Shafer [SHA 76], but the theory's origins go back to Dempster [DEM 68], via his work on lower and upper bounds in a probability distribution family. Using the mathematical formalism developed, Shafer presented the use of belief functions for modeling uncertain knowledge [BAL 97]. Indeed, this theory enables situations of ignorance to be managed. In this part, we begin by introducing evidence theory and its different mass functions, then we tackle evidence combination in the theory and we describe the decision rule for taking a good combination. Then we will present our work in order to see how our method performs in multi-target tracking and data combination, ending this chapter with a conclusion.

10.2. Dempster–Shafer theory

The Dempster–Shafer theory (DST), or evidence theory, is often considered to be a generalization of Bayesian probability theory [DEM 68, SMA 04]. The theory was introduced via the notions of greater and lesser probabilities by Dempster [DEM 67]. It was then refined and presented by Shafer through belief and plausibility functions [SHA 76].

The applications only concern sets of discreet definitions at credal and pignistic level. Based on a mass distribution of evidence m defined across all the propositions Θ (the space of finite cardinal possible worlds), it links degrees of belief and plausibility to parts of A (a group of propositions or information) of Θ.

10.2.1. *Frame of discernment*

DST requires a definition of a framework of discernment written Θ, made up of a set of all the hypotheses required for a complete description of a present situation. The hypotheses in the set are assumed to be mutually exclusive. Also, the set is considered to be exhaustive (closed world assumption). Generally, the frame of discernment Θ defines the hypotheses for which the information sources provide confidence. For a set including HN simple possible hypotheses in a precise context, the frame of discernment can be defined as follows:

$$\Theta = \{H1, H2, \ldots\ldots, Hn, \ldots\ldots, HN\} \quad [10.1]$$

The reasoning framework used in DST is the basic reasoning framework, written 2Θ, which is defined as being the set of all the possible sub-sets taken from simple hypotheses from a frame of discernment Θ which permits the combination. The reasoning framework 2Θ is designed as being the power set, which is an extended version of the frame of discernment where it contains not only simple hypotheses (singletons), but also any combination of singletons (composite hypotheses) [DJI 08]:

$$2\Theta = \{\emptyset, H1, H2, \ldots, HN, H1 \cup H2, \ldots, H1 \cup H2\ldots HN\} \quad [10.2]$$

Thus, in set 2Θ we find: the empty set (0), the simple hypotheses H1, and composite hypotheses obtained by combining the simple hypotheses, which represent unknowns or the inability to differentiate different simple hypotheses.

10.2.2. *Elementary mass function*

If Θ is a frame of discernment, we defined a mass function m as:

$$m : 2^{\Theta} \rightarrow [0,1] \quad [10.3]$$

This function checks the following properties:

$$m(\varnothing) = 0$$

$$m(A) \in [0, 1] \qquad [10.4]$$

$$\sum_{A \subseteq \Theta} m(A) = 1$$

A set's elementary mass function is used to express a degree of confidence (level of belief) accorded to this set. In DST, the mass function m is defined for each set A of the reasoning framework 2Θ, $A \in 2^{\Theta}$ answering two demands with respect to the closed-world hypothesis: that the mass of the empty set should be null and the sum of the masses of all the possible sets A is equal to 1.

The quantity m(A) is the part of belief placed strictly on A without being placed on the hypotheses of which it is made up. The quantity is differentiated from a probability by the fact that all the belief is distributed not only on the singleton hypotheses Hn, but also on the composite hypotheses A. We can therefore accord a proposition a part of belief and thus assign to all hypotheses contained in the proposition, a belief in the realization of each of them without choosing one precisely.

Here we show some mass functions in particular that will be used in information modeling [LEF 01]:

– focal element;

– F symbolizes the set of focal elements of the mass function m:

$$F = \{A \in 2^{\Theta} \mid m(A) > 0\}; \qquad [10.5]$$

– a focal element is a sub-set A of Θ whose mass m (A) is non null;

– total ignorance function [LEF 01];

defined by:

$$m(\Theta) = 1 \qquad [10.6]$$

This is the case of indetermination or total ignorance. The observer knows that the assumption solution is found in the frame of discernment, but nothing more can be said about it. It cannot divide its knowledge on a set smaller than Θ.

Total uncertainty function:

defined by:

$$m(H_n)=1 \qquad [10.7]$$

Division of the belief mass is concentrated on a singleton hypothesis. The observer is confident of knowing the assumption solution, and models their knowledge by a total belief mass in hypothesis Hn.

10.2.3. *Credibility and plausibility functions*

In the Bayesian approach, the uncertainty concerning an event is measured by a single value: the probability and imprecision on the uncertainty measurement are assumed to be null. In DST, set A's belief value (credibility) can be interpreted as the minimal uncertainty value on A, and its plausibility value can be interpreted as the maximal uncertainty value on A. Therefore, the uncertainty on A is represented by interval values [Bel(A), Pl (A)], which is called the confidence interval, in which Bel represents the belief or credibility function and Pl the plausibility function. The length of this interval quantifies the imprecision concerning the uncertainty of the source on set A.

10.2.3.1. *The credibility function*

The belief accorded to a sub-set A of Θ represents the total confidence from which the sub-set A can be accorded. The credibility function, Bel: $2^{\Theta} \rightarrow [0,1]$, can be defined using its mass function by:

$$Bel(A)=\sum_{B\subseteq A} m(B) \qquad \forall A\subseteq\Theta \qquad [10.8]$$

Bel(A) represents the entirety of the belief accorded to the elements of the disjunction of the A hypotheses.

10.2.3.2. *Plausibility function*

The plausibility function, Pl: $2^{\Theta} \rightarrow [0,1]$, represents the upper limit up to which a sub-set A of Θ can be considered credible, and it can be defined using the mass function:

$$Pl(A)=\sum_{(A\cap B)\neq\varnothing} m(B) \qquad \forall A\subseteq\Theta \qquad [10.9]$$

Pl (A) is interpreted as the part of belief that could potentially be allocated to A, that is to say all the information contained in the sub-sets that have a non–null intersection with A.

10.2.3.3. *Relationship between the credibility function and the plausibility function*

A's plausibility is generally linked to the credibility of A's complement. It corresponds to all the information that does not credit the truth of A's complement.

$$Pl(A) = 1 - Bel(\overline{A}) \qquad [10.10]$$

In the Dempster–Shafer theory, the interval defined by [Bel (A), Pl (A)] can be interpreted as a frame of A's real probability, Bel (A) is "low probability" and Pl (A) is "low probability". The belief mass distribution m and the credibility function Bel are two equivalent representations of the same information.

10.2.3.4. *Other measures of mass function*

Using the mass allocation, we can cite other measures such as the communality function or interchangeability function, Q: $2^{\Theta} \rightarrow [0,1]$ can be defined using the mass function by:

$$Q(A) = \sum_{A \subseteq B} m(B) \qquad \forall A \subseteq \Theta \qquad [10.11]$$

There is a bijection between functions m, Bel, Pl and Q as shown on Figure 10.1, which contains exactly the same information. In particular, the mass function is found using a belief function via the Möbius transformation, defined by:

$$m(A) = \sum_{B \subseteq A} (-1)^{|A-B|} Bel(B) \qquad [10.12]$$

$|A - B|$ represents the number of elements of A that do not form part of B.

10.2.3.5. *Interpreting plausibility and credibility functions*

We can represent these two measurements (credibility and plausibility) using Figure 10.2. This figure explains the minimal and maximal likelihood labels to which we sometimes allocate credibility and plausibility in the context of evidence theory. In this figure, we visualize the fact that credibility regroups all the masses of focal elements (B1, B2) included in the sub-set A, whereas the plausibility corresponds to all the intersecting masses (B1, B2 and B3) with the sub-set considered, A.

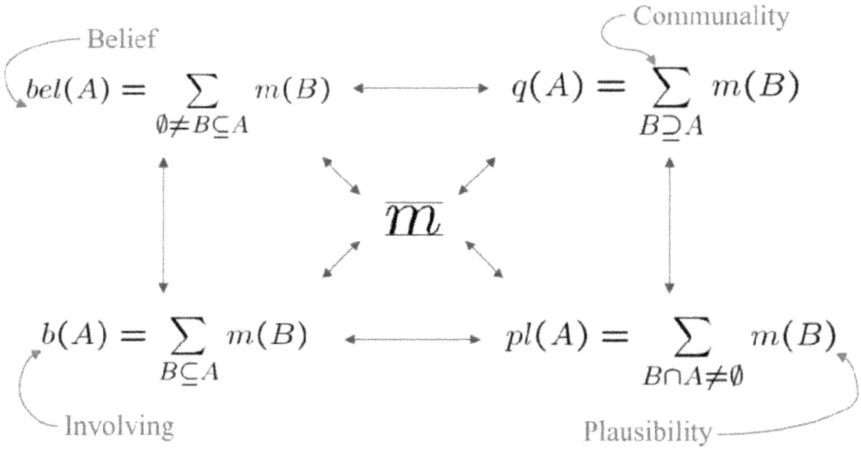

Figure 10.1. *Different mass function measures*

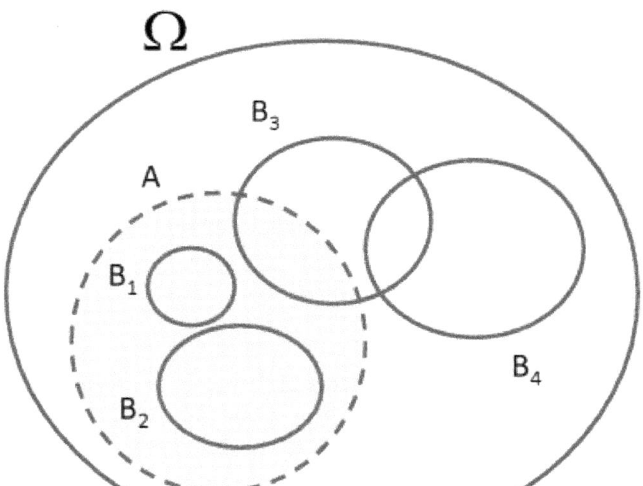

Figure 10.2. *Credibility and plausibility of a set A*

In this framework, Bel (A) and Pl (A) can be seen, respectively, as a low probability and a high probability [SHA 76]. It should be noted that there are other symmetrical credibility and plausibility functions, that is disbelief $Pl(\overline{A})$ (credibility of the contrary) and doubt $BL(\overline{A})$ (the quantity of belief mass remaining). The existing relationships between these different measures are represented in Figure 10.3.

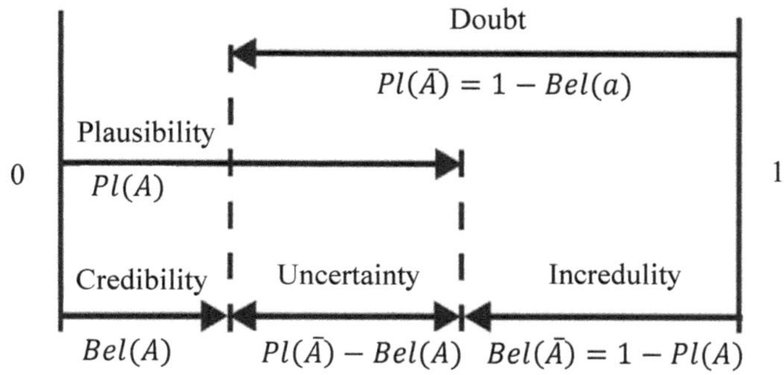

Figure 10.3. *Representing the evidence of a set A*

10.3. Rules of combination

The Dempster–Shafer theory enables information taken from different sources to be combined, according to Shafer, if two belief functions are defined on a single frame of discernment such that this frame of discernment distinguishes the pertinent interaction between two pieces of information. The combination of evidence is involved when new information coded in BF form is available, which must be fused with the already existing BF. The information fusion is presented as a solution enabling more reliable information to be accessed, or a synthesis of knowledge about a multi-source environment.

10.3.1. *Basic rules of combination*

Suppose S1 and S2 are two independent sources of information. These sources are assumed to be totally reliable. The information obtained via these sources is modeled with the help of belief functions called m_1 and m_2 respectively. We write the belief function resulting from the conjunctive combination of m_1 and m_2 as $m_{\cap}$:

$$m_{\cap} = m_1 \cap m_2 \quad [10.13]$$

10.3.1.1. *Conjunctive combination*

Conjunctive combination is an elementary rule of combination based on the intersection operator. It consists of taking the sum of all the possibilities of conjunction between all the sources [SME 90].

$$m_{\cap}(A) = \sum_{B \cap C = A} m_1(B) m_2(C) \qquad \forall A \subseteq \Theta \tag{10.14}$$

We can generalize this rule to J information sources producing J BF. By writing $m_{\cap} = \cap_{j=1}^{J} m_j$, we will have:

$$m_{\cap}(A) = \sum_{A_1 \cap \cdots \cap A_J = A} \left(\prod_{j=1}^{J} m_j(A_j) \right) \qquad \forall A \subseteq \Theta \tag{10.15}$$

10.3.1.2. *Disjunctive combination*

Disjunctive combination is an elementary rule of combination based on the union operator. It consists of taking the sum of all the disjunction possibilities between all the sources. For the combination of J sources, this rule is given by the following equation:

$$m_{\cup}(A) = \sum_{A_1 \cup \cdots \cup A_J = A} \left(\prod_{j=1}^{J} m_j(A_j) \right) \qquad \forall A \subseteq \Theta \tag{10.16}$$

In a general way, the basic combination rule produces a non-normalized mass function, that is to say that $m\cap(\varnothing) \neq 0$. This hypothesis is not possible in the closed-world scenario. It is therefore necessary to introduce a normalization stage. This normalized conjunctive combination law is better known by the name of Dempster's rule of combination.

10.3.2. *Dempster's rule of combination*

The Dempster–Shafer theory (DST) assumes that the information sources are mathematically independent. Dempster's rule of combination is in reality a normalized conjunctive combination functioning on the power set 2Θ.

It works under the closed world constraint where the mass of the empty set is thought to be null. Consequently, if the conjunctive rule generates a non-null mass for the empty set, this mass is attributed to the mass of conflicts. The Dempster rule, also called the orthogonal sum, of two belief functions m_1 and m_2 is written in the following way:

$$m_{\oplus} = m_1 \oplus m_2 \qquad [10.17]$$

The Dempster rule redistributes this mass of conflicts uniformly on the other focal elements as follows:

$$m_{\oplus}(A) = \frac{1}{1-K} \sum_{B \cap C = A} m_1(B) m_2(C) \qquad \forall A \subseteq \Theta, A \neq \varnothing \qquad [10.18]$$

where K is defined by:

$$K = \sum_{B \cap C = \varnothing} m_1(B) m_2(C) \qquad [10.19]$$

We call the degree of conflict (or simply conflict) between the two mass functions m_1 and m_2 written K, where K reflects the conflictual belief mass existing between the two masses to be combined. This mass varies in the interval [0, 1] according to functions m_1 and m_2. When K is null, the sources are in perfect agreement. On the contrary, when this mass is equal to 1, the sources are in total conflict and the information cannot be fused.

We can generalize the Dempster combination to J belief functions by writing $m_{\oplus} = \oplus_{j=1}^{J} m_j$. the function m $\otimes$ is therefore defined in the following way [LEF 01]:

$$m_{\oplus}(A) = \frac{1}{1-K} \sum_{A_1 \cap \cdots \cap A_J = A} \prod_{j=1}^{J} \left(m_j(A_j) \right) \qquad \forall A \subseteq \Theta, A \neq \varnothing \qquad [10.20]$$

$$K = \sum_{A_1 \cap \cdots \cap A_J = \varnothing} \left(\prod_{j=1}^{J} m_j(A_j) \right) \qquad [10.21]$$

Dempster's rule of combination is commutative and associative. Consequently, the information's order of combination should not influence the results when the Dempster rule is applied.

Finally, Dempster's rule of combination has been the subject of much criticism. Most of the criticisms are linked to normalization. Smets [SME 90] even suggests abandoning this normalization.

10.4. Decision rules

Once the resulting mass m is obtained, the decision can be taken and different decision rules can be defined, the most popular being the maximum plausibility rule and the maximum credibility rule. Using these functions, a probability function called pianistic probability is defined [BAR 88, BAR 00]. Generally, the decision function δ is defined for an incoming form X to be classed by:

$$\delta(X) = Hn \quad \text{or} \quad Hn = \text{Arg}\,[\max \gamma(Hi)],\ Hi \in \Theta \qquad [10.22]$$

where γ is the credibility function (Bel), or plausibility function (pel) or pianistic probability.

10.5. Digital simulation

In this chapter we will show DST performance in multi-target tracking in different cases and essentially the probability of a good data combination, as DST is known by the fusion of uncertain data (Dempster combination); this property of the method will be used to link the corresponding measure to each target as Figure 10.4 indicates, where Θ represents the set of targets and Ω represents the set of measurements.

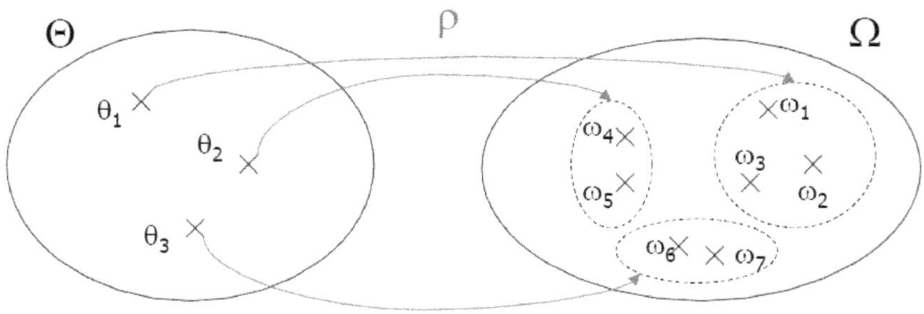

Figure 10.4. *Linking measurements to the corresponding targets*

10.5.1. *Movement models*

In our work, we will use a rectilinear movement model at constant velocity for different cases:

– three parallel targets, separated by 50 m, 100 m and 150 m;

– three targets that increase to N /2.

Suppose that at instant k we have the following vector (state model):

$$\underline{X}_k = \left[x_k, \dot{x}_k, y_k, \dot{y}_k\right]^T \qquad [10.23]$$

The latter is made up of the target's parameters (abscissa, abscissa velocity, ordinate and velocity on the ordinate) which are unknown and should be estimated.

In the rectilinear movement model at constant velocity, the function $f_l\ (X_k)$ and the matrix G_1 are given by [GAN 05]:

$$f_1\left(\underline{X}_k\right) = \begin{bmatrix} 1 & T & 0 & 0 \\ 0 & 1 & 0 & 0 \\ 0 & 0 & 1 & T \\ 0 & 0 & 0 & 1 \end{bmatrix} \underline{X}_k \text{ And } G_1 = \begin{bmatrix} T^2/2 & 0 \\ T & 0 \\ 0 & T^2/2 \\ 0 & T \end{bmatrix} \qquad [10.24]$$

where T represents the sampling period (a radar tower period).

10.5.2. *Working model*

The method adapted for our work is represented on Figure 10.5, where we will model the information by the elementary mass functions, then combine the resulting masses with the Dempster combination, and take the maximum of the credibilities as the decision rule.

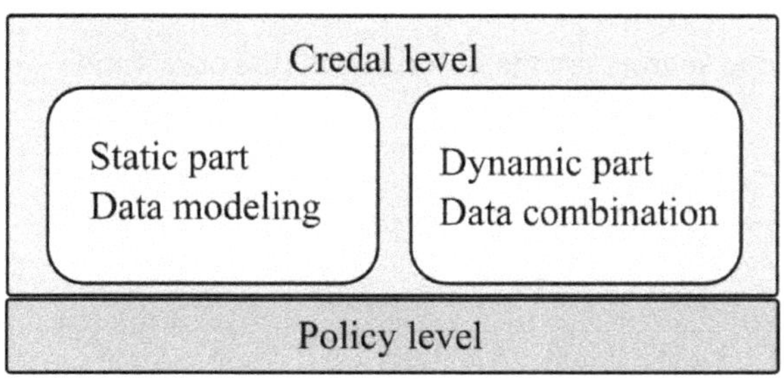

Figure 10.5. *Association steps*

10.5.2.1. *Static part:*

In the framework of evidence theory for data combination, there are no generic methods for obtaining the mass function. In most cases, the modeling used depends on the application envisaged; in our work, we have adopted a model for combining data in the context of multi-target pursuit, inspired by Appriou's model [APP 91].

To construct our model, we use the following stages:

– our frame of discernment is made up of two hypotheses, $\Omega_{ij} = \{H_1, H_2\}$ with:

- H_1: the target i is linked to the observation j',

- H_2: the target i is not linked to the observation j, so $H_2 = \overline{H_1}$;

– we define the set 2N (N=2) part of Ω_{ij} by:

$$2^{\Omega_{ij}} = \{\{H_1\}, \{H_2\}, \{H_1, H_2\}, \varnothing\};$$

– we accord each element a belief mass $A \in 2^{\Omega_{ij}}$:

- $m_{ij}(\{H_1\}) = m(i,j)$: the mass of association of the target *i* to the observation *j*,

- $m_{ij}(\{H_2\}) = 1 - m(i,j)$: the mass of ignorance of association of the target *i* to the observations *j'* *(j'≠j)*,

- $m_{ij}(\{H_1, H_2\}) = 0$: the mass of ignorance of association of the target *i* to the observation *j*,

- $m_{ij}(\varnothing) = 0$: the mass of association and non-association of the target *i* to the observation *j*;

– our information sources are the observations (the observation's dimension = 2), which is:

$$G(i,j) = \frac{1}{2.\pi\sqrt{|S_i|}}.\exp\left(-\frac{1}{2}.(y_j - yp_i)^T.(S_i)^{-1}.(y_j - yp_i)\right) \qquad [10.25]$$

We define m*(i,j)* by [KOU 09, JAN 96]:

$$m_{norm}(i,j) = \frac{G(i,j)}{max_j[G(i,j)]} \rightarrow m(i,j) = \frac{m_{norm}(i,j)}{\sum_j^{nof\ measures}[m_{norm}(i,j)]} \qquad [10.26]$$

– no. of measurements: number of observations received at instant *k*;

– y_j: jth observation received;

– y_{pi}: the predicted observation of the target i;

– if: innovation covariance matrix of the target i;

The model becomes:

$$\begin{cases} m_{ij}\left(\{H_1\}\right) = m(i,j) \\ m_{ij}\left(\{H_2\}\right) = 1 - m(i,j) \\ m_{ij}\left(ign\right) = 0 \\ m_{ij}\left(\varnothing\right) = 0 \end{cases} \quad [10.27]$$

where

$$\left\{ \begin{array}{c} \mathrm{m(H_1)association\ probability\ of\ the\ target}\ i\ \mathrm{to\ the\ observation}\ j \\ \mathrm{m(H_2)\ non\ association\ probability\ of\ the\ target}\ i\ \mathrm{to\ the\ observation}\ j \\ \mathrm{m(ign)ignorance\ probability\ on\ the\ association\ of\ the\ target}\ i\ \mathrm{to\ the\ observation}\ j \end{array} \right\}$$

10.5.2.2. *Dynamic part*

In this stage we apply the conjunctive combination rule [10.14] for each target and each measurement, we will then have the mass functions table (Table 10.1). Afterwards, we will normalize in relation to Ki [10.19], and we will therefore have the normalized evidence.

	A1	A2	...	AJ	$\varnothing$
P1	$m_{\cap 11}$	$m_{\cap 12}$	...	$m_{\cap 1J}$	$m_{\cap 1\varnothing}$
P2	$m_{\cap 21}$	$m_{\cap 22}$	...	$m_{\cap 2J}$	$m_{\cap 2\varnothing}$
...	...	...	...	...	...
PI	$m_{\cap I1}$	$m_{\cap I2}$	...	$m_{\cap IJ}$	$m_{\cap I\varnothing}$

Table 10.1. *Table of combined masses*

To link the measurements to each corresponding target, we apply the decision rule [10.21]. For each target, we choose the observation with the maximum credibility on the mass after combination to reveal the state of the target.

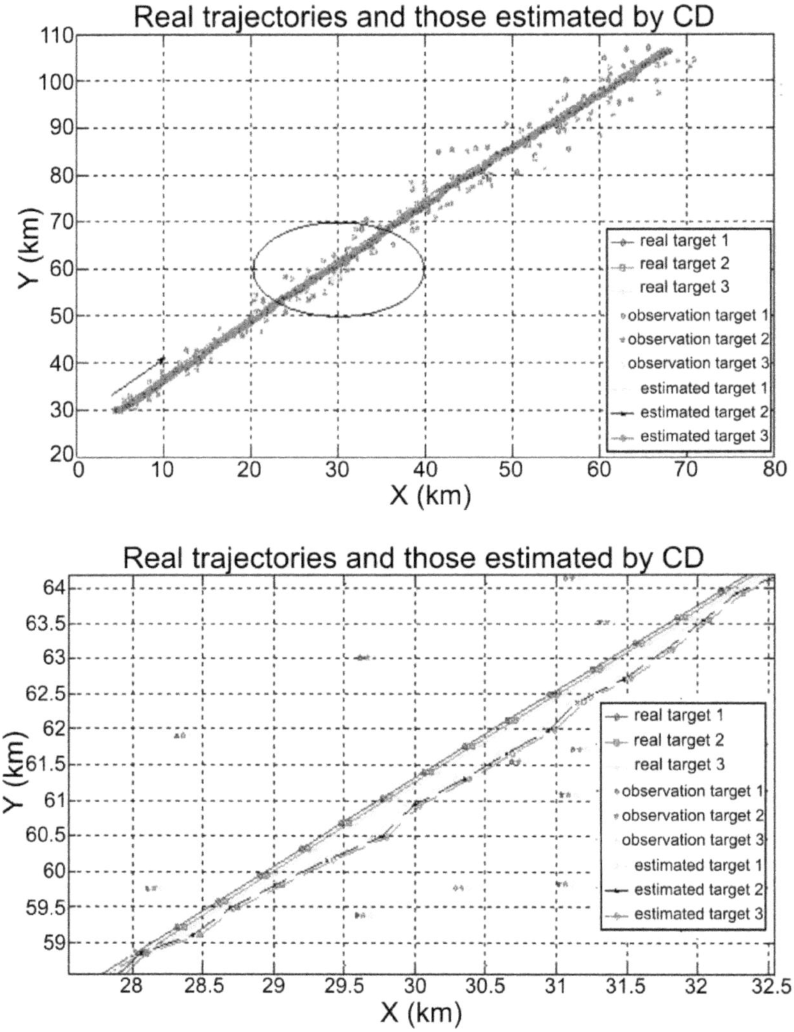

Figure 10.6. *Real trajectories and those estimated using DST for three parallel targets. For a color version of this figure, see www.iste.co.uk/benslama/quantum.zip*

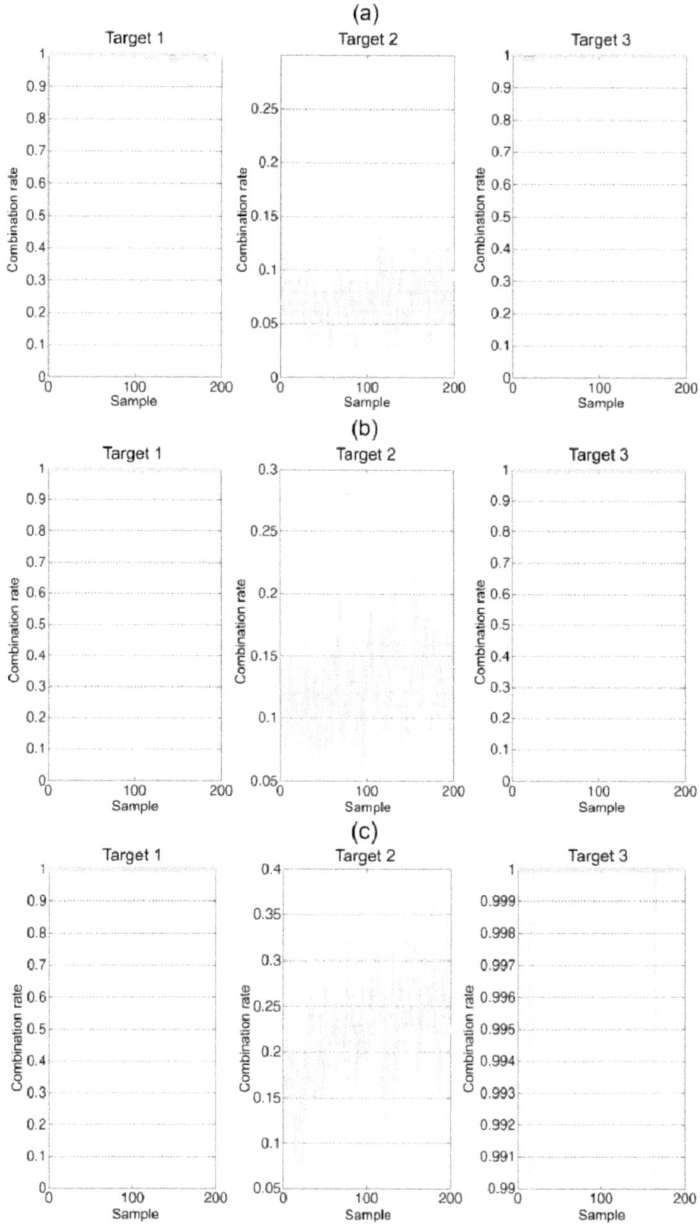

Figure 10.7. *Combination rate for the three parallel targets; a) separations between the targets 50 m; b) separations between the targets 100 m; c) separations between the targets 150 m*

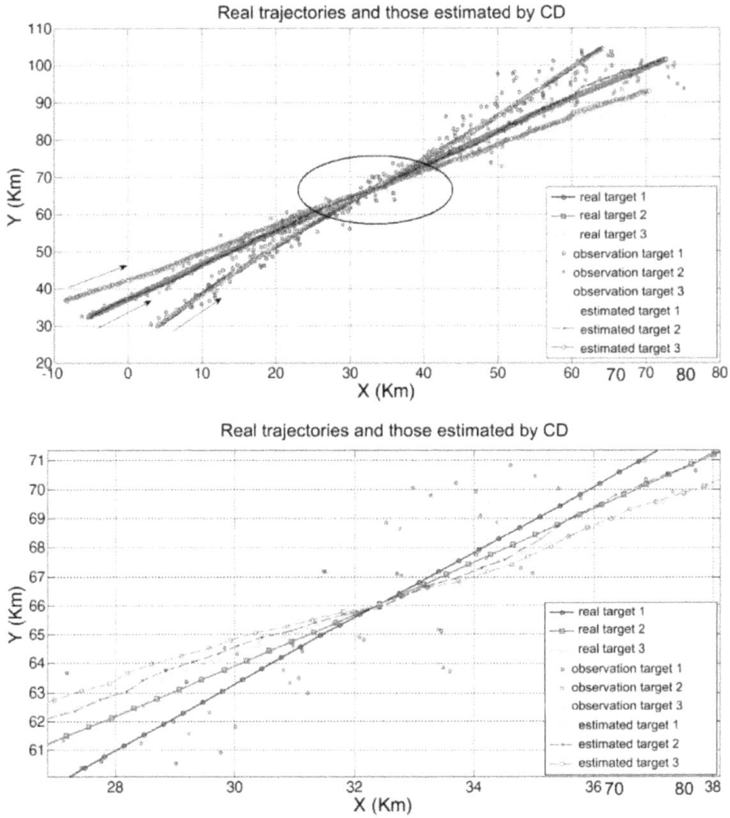

Figure 10.8. *Real trajectories and those estimated using DST for three intersecting targets. For a color version of this figure, see www.iste.co.uk/benslama/quantum.zip*

The results obtained enable the following observations to be made.

10.5.3. *3 parallel targets*

The trajectories estimated by DST are very close to the real trajectories: the separation between the targets is clear and it can be seen that the three targets are not confused or intersecting.

The probability of a good combination is very high except for the middle target and this is due to the separation between the targets, if we increase the separation between the targets, the probability of a good combination increases.

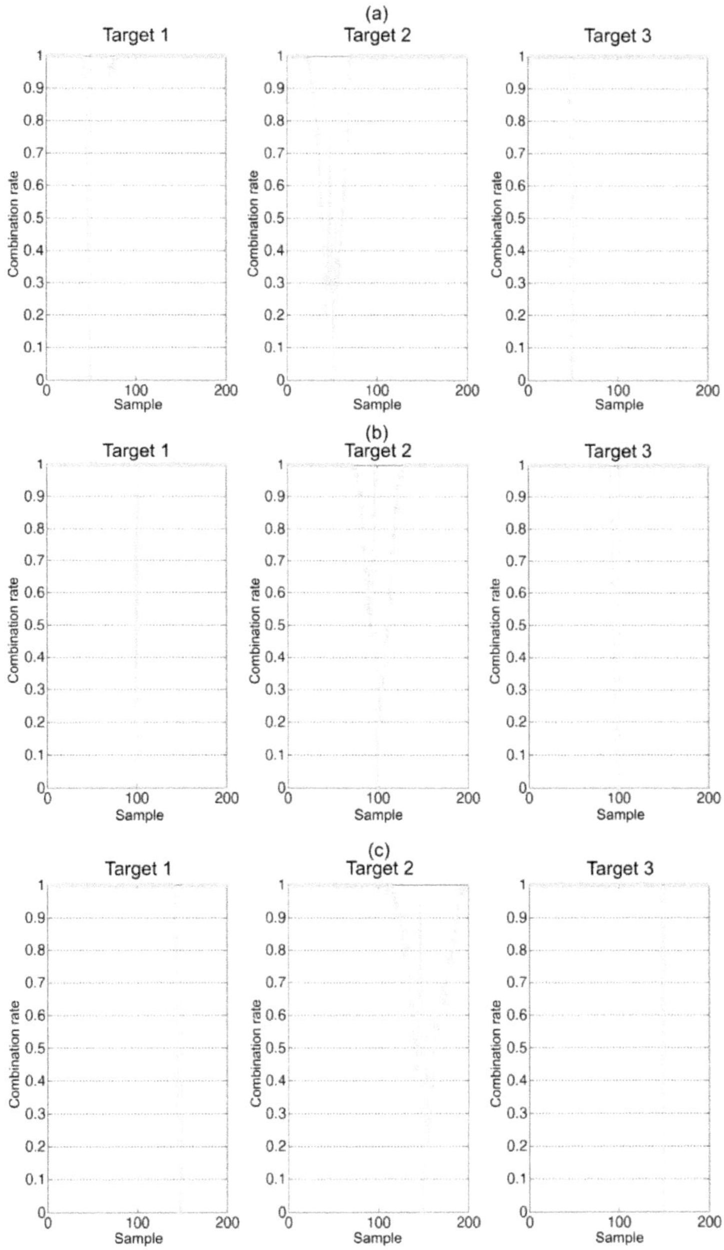

Figure 10.9. *Combination rate for three intersecting targets; a) Intersection point N/4; b) Intersection point N/2; c) Intersection point 3N/4*

10.5.4. *3 intersecting targets*

The trajectory estimated by DST is similar to the real trajectories for the three targets: the separations between the targets are clear and the point of the targets' intersection is very close to the real one.

The probability of a good combination is very high unless it falls at the points of intersection (N/4, N/2, 3N/4) and it resumes just after the intersection except at target 2 and this is due to the separation between the targets.

10.6. Conclusion

In our work, we have presented the Dempster–Shafer theory, from which evidence theory is taken. Using probabilistic mathematics, we have presented a formal framework for reasoning in uncertainty and a model that enables knowledge to be modeled. In fact, via the mass functions that are the measuring tools for subjective probability, it is possible to evaluate the degree of truth in an expert opinion. With the introduction of evidence masses, the combination rule enables information to be processed so as to result in reliability. This greatly helps decision-making based on maximum credibility.

Data modeling determines association performances; in this chapter we work with a modeling model inspired by Appirou's work [APP 91].

In our work, we have seen performances for multi-target tracking and for our method's association, the results obtained using simulation have demonstrated a good performance for the Dempster–Shafer theory.

11

InSAR Radar

11.1. Introduction

In recent years, a great deal of interest has been given to interferometric radar with synthetic aperture (InSAR), with which two or more complex images with the same area on land are taken into consideration. It is necessary to specify that these images are taken at different times and/or in different orbits. They can be used to calculate each interferogram, the samples of the coherence function of the scene or resolved images [IGL15, GAO 15, SIC 15].

For these applications, the SAR image's phase is a delicate characteristic and should be monitored in the image processing. The precision of the SAR interferograms' usable phase is affected by decorrelation due to:

– decorrelation of the temporal scene (change in surfaces, water or vegetation) [GAB 88];

– noise from the radar's receiver;

– aberrations in the phase introduced during data acquisition or processing [LI 90];

– the transfer functions' poor spectral alignment caused by different aspect angles [HAG 70];

– the statistics of the inferograms' phase from these last effects are studied.

11.2. Coherence

The degree of coherence between two complex data sets Z_1 and Z_2 is defined as the correlation coefficient [ZEB 86].

$$\gamma = \frac{E\{Z_1 Z_2^*\}}{\sqrt{E\{|Z_1|^2\}.E\{|Z_2|^2\}}} \qquad [11.1]$$

According to the decorrelation mentioned in the introduction, ε is composed of the following factors.

$$\gamma = \gamma_{scene} \cdot \gamma_N \cdot \gamma_H \qquad [11.2]$$

where γ_{scene} indicates the long term correct coherence, the index N for the noise and γ_H summarizes the transfer functions' influence, describing the SAR imaging process including the decorrelation base and the processor's differential aberrations.

11.3. System model

The interferometric phase is made up of the phase caused by the geometry which carries the useful information and the trouble. We are interested in a quantitative estimation of the trouble introduced using the data acquisition and processing. From there, it is supposed that the geometric part of the phase has been reversed by an appropriate compensation and $\gamma_{scene} = 1$.

In this case, the undisturbed interferogram phase will be null throughout.

Then the system model from Figure 11.1 is applicable.

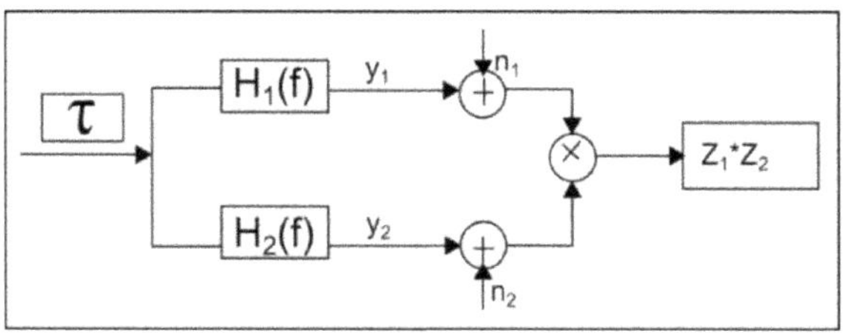

Figure 11.1. *Formation system model INSAR [HAG 70]*

The input signal, the scene's reflection, is introduced by the complex circular white stationary Gaussian process, x of the spectral density of the power $2\sigma^2$. The transfer functions H_1, H_2 that describe the two imaging processes are generally different, because of the difference in the orbits, and the differential processor's possible aberrations. The two filters' output signals y_1 and y_2 form a complex stationary Gaussian process. The terminal noise's different processes n_1 and n_2 are taken into account in the main inspections SNR_1 and SNR_2, (signal to noise ratio).

The interferogram is defined by $Z_1{*}Z_2$ and the phase by: $\Delta\varphi = \arg\{Z_1.Z_2\}$.

The autocorrelation functions of x and n_1 are respectively:

$$R\xi\xi(\tau) = 2\sigma^2\delta(\tau) \qquad [11.3]$$

$$R_{n1n1}(\beta) = 2\sigma_1^2 \cap \left|H_{1P}(f)\right|^2 \exp\{-2\pi jf\beta\}df \qquad [11.4]$$

where $H_{1P:}$ represents the process's transfer function in imaging branch # 1.

The signal to noise ratio or exit system Z_1 is therefore:

$$SNR_1 = \frac{\sigma^2 \int_{-\infty}^{+\infty} \left|H_1(f)\right|^2 df}{\sigma_1^2 \int_{-\infty}^{+\infty} \left|H_{1P}(f)\right|^2 df} \qquad [11.5]$$

11.4. Inferometric phase statistics

After calculating the probability density function $pdf_0(\phi)$ and the interferometric phase, we obtain [MAD 99]:

$$\mathrm{Pdf}(\Delta\phi) = \mathrm{pdf}(\Delta\phi - \phi_0) \qquad [11.6]$$

with:

$$\phi_0 = \arg\{\gamma\} \qquad [11.7]$$

$$\mathrm{pdf}\,(\phi) = \frac{1-|\gamma|^2}{2\pi}\,\frac{1}{1-|\gamma|^2\cos^2\phi} \times \left\{1+\frac{|\gamma|\cos\phi\arccos(-|\gamma|\cos\phi)}{\sqrt{1-|\gamma|^2\cos^2\phi}}\right\} \qquad [11.8]$$

This probability density function is known generally in the context of higher order statistics. Recently, it has been used in polarimetry. Assuming that $pdf_0(\phi)$ is defined in the interval $[-\pi, \pi]$, then:

$$E(\phi) = \int_{-\pi}^{+\pi} \phi.pdf_0\,(\phi)\mathrm{df} = 0 \qquad [11.9]$$

From there, the biased stage is:

$$E(\Delta\phi) = \phi_0 = \arg(\gamma) \qquad [11.10]$$

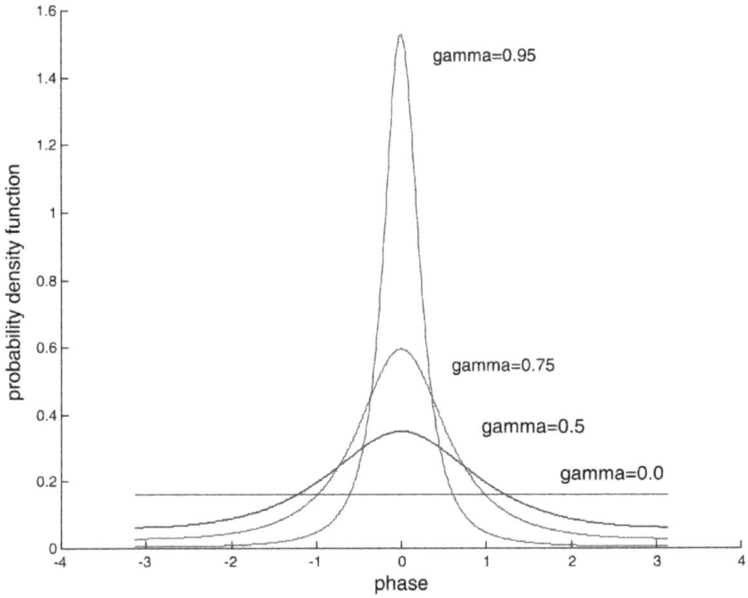

Figure 11.2. *The inferometric phase's probability density function for different values of the correlation coefficient*

In Figure 11.2, we see how the probability density function varies according to the phase ϕ for four values of $|\gamma| = |\gamma_N| \times |\gamma_H|$:

– the shape of the pdf(ϕ) is a Gaussian form;

– the pdf's maximum (ϕ) is for $\phi = 0$ for the four values of $|\gamma|$.

Therefore $\Delta\phi = \phi_0 \Rightarrow \arg\left[Z_1 Z_2^*\right] = \arg(\gamma)$

The inferometric phase is undisturbed for this value of $\phi = 0$:

– the increase of $|\gamma|$, gives us a narrowing of the shape of the pdf (ϕ);

– it is noted that ϕ_0 changes the probability density function.

The phase's variance:

$$E\{(\Delta\phi - \phi_0)^2\} = pdf_0(\phi)d\phi = \sigma_\phi^2 \quad [11.11]$$

This last interval has been digitally evaluated for the examples given.

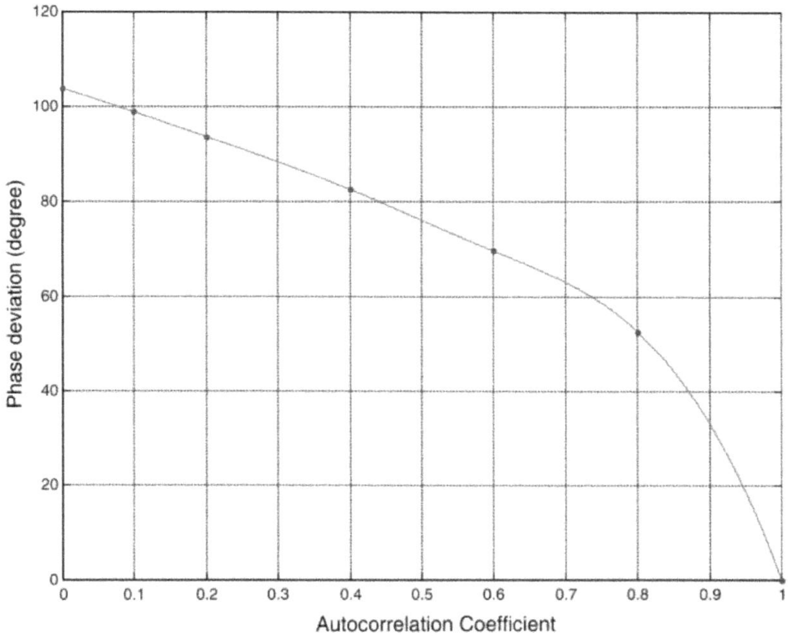

Figure 11.3. *Standard deviation of the interferometric phase compared to the size of the correlation coefficient*

Figure 11.3 shows the deviation of the standard phase σ_ϕ compared to $|\gamma|$.

– the infinity of the slope can be written $|\gamma|$ = 1;

– the phase deviation decreases as $|\gamma|$ increases;

– the phase deviation σ_φ is equal to 103° for a $|\gamma| = 0$;

– note that the bias phase $\phi_0 = \arg(\gamma)$ does not change the variance.

The correlation coefficients γ_N and γ_H can be expressed in terms of SNR_1, SNR_2 and the transfer functions H_1, H_2:

$$\gamma_N = |\gamma_N| = \frac{1}{\sqrt{\left(1+SNR_1^{-1}\right)\left(1+SNR_2^{-1}\right)}} \qquad [11.12]$$

and

$$\gamma_H = \frac{\int_{-\infty}^{+\infty} H_1(f).H_2(f)^* \, df}{\sqrt{\int_{-\infty}^{+\infty} |H_1(f)|^2 . |H_2(f)|^2 \, df}} \qquad [11.13]$$

The results are summarized as follows:

– the mean value φ_0 of the inferometric phase is equal to the phase of the complex correlation coefficient correlation γ which can be expressed in turn as a functional of the transfer functions and does not depend on SNR, the variance σ_φ^2 depends only on the size of the correlation coefficient and can be determined by the data of H_1, H_2 and SNR.

11.5. Quantitative examples

The effects of the specific transfer function and the end noise on the inferometric phase statistics will be evaluated quantitatively [BAM 98].

11.5.1. *The final noise*

In the case of identifying transfer functions $H_1 = H_2$, γ is determined only by [11.12].

If the two inspections' signal to noise ratios are identical, γ is reduced to:

$$\gamma = \frac{1}{1+SNR^{-1}}$$

γ is known according to [11.13] and [11.14].

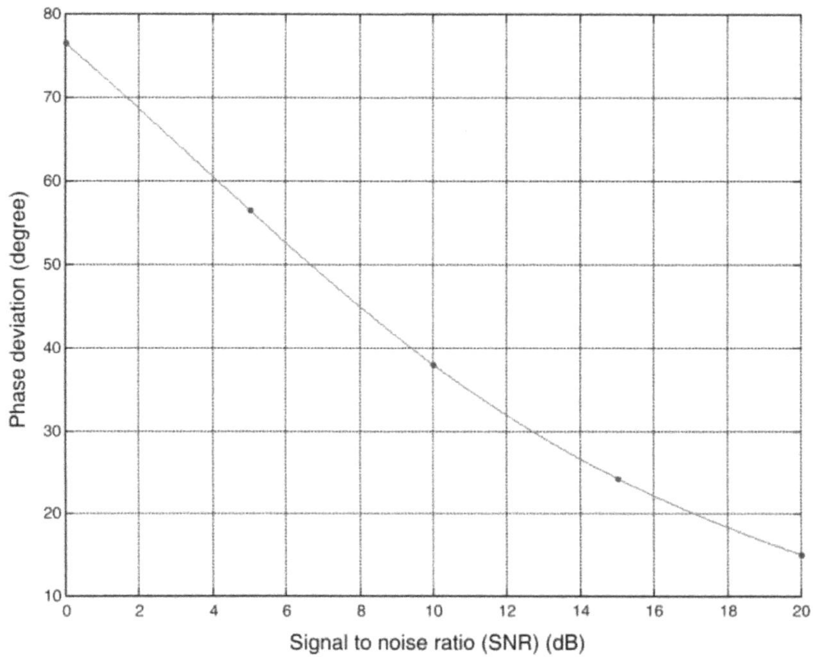

Figure 11.4. *Standard deviation for the phase compared to the SNR*

Figure 11.4 shows the signal to noise ratio's influence on the phase deviation.

When the SNR increases, the phase deviation decreases.

The SNR ratio increases, we see a weak noise leading to a clearer signal, and so a small deviation or disturbance gives us a more correct phase difference, and so a more exact third dimension:

– for an SNR equal to 20dB, the phase is equal to 12°;

– the correlation coefficient varies proportionally with the SNR.

11.5.2. *The phase aberrations*

We concentrate on the effects of aberrations in the imaging process. It is assumed that the spectral envelopes of H_1 and H_2 are identical, however its phases are different. H_1 can be presented as a real value and the phase difference is introduced, which can be called "differential phase aberrations" in H_2. For the examples given in this section, H_1 and H_2 are low-pass filters:

$$\mathrm{H}_1(\mu,\nu)=\mathrm{rect}\left(\frac{\mu}{\mathrm{B}_\mu}\right).\mathrm{rect}\left(\frac{\nu}{\mathrm{B}_\nu}\right) \quad [11.15]$$

$$\mathrm{H}_2(\mu,\nu)=\mathrm{H}_1(\mu,\nu).\exp\{-\mathrm{j}\,\psi(\mu,\nu)\} \quad [11.16]$$

or

μ: Azimuth frequency

ν: Gate frequency

Bμ: Size of the azimuth band

Bν: Size of the gate band

and ψ (μ,ν) represents the phase's differential aberration phase between the two imaging processes. From there, the complex correlation coefficient can be expressed by:

$$\gamma=\frac{1}{\mathrm{B}_\mu\mathrm{B}_\nu}\int_{-\mathrm{B}_\nu/2}^{+\mathrm{B}_\nu/2}\int_{-\mathrm{B}_\mu/2}^{+\mathrm{B}_\mu/2}\exp\{\mathrm{j}\,\psi\}\mathrm{d}\mu\mathrm{d}\nu \quad [11.17]$$

the aberrations of which the Taylor series is developed are fairly small, so ψ (μ,ν) can be writing as:

$$\psi(\mu,\nu)=2\pi\{\psi_{00}+\psi_{10}.\mu+\psi_{01}.\nu+\psi_{11}.\mu\nu+\psi_{20}.\mu^2+\psi_{02}.\nu^2+\psi_{21}.\mu^2\nu+.....\} \quad [11.18]$$

ψ_{00}: constant phase error (not considered in our study).

ψ_{10}: poor geometric registration in azimuth

ψ_{01}: poor geometric registration in the gate.

ψ_{20}: divergence in the azimuth.

ψ_{00}: divergence in the gate

ψ_{11}: migration of the uncompensated linear distance.

ψ_{21}: migration of the uncompensated quadratic distance.

$|\gamma|$: the bias phase, and the variance for each individual phase aberration will be evaluated.

ψ_{01} or ψ_{10}: poor relative geometric registration in the azimuth or the gate.

It is possible that an error in registrating the two complex images can be presented as a linear phase aberration (only poor registration in the azimuth is considered).

If α represents the relative change between the two images in fractions of a resolution element $1/\beta_\mu$:

$\psi_{10} = 1/\beta_\mu$,

then we obtain:

$$\psi(\mu,\nu) = 2\pi.\psi_{10}.\mu = 2\pi.\alpha.\mu/B_\mu \qquad [11.19]$$

and

$$\varphi_0 = 0 \text{ and } |\gamma| = \text{sinc}(\alpha) \qquad [11.20]$$

Therefore, the poor registration does not introduce a bias phase but the phase's variance. Figure 11.5 shows the phase's standard deviation compared to the change a. For example, the poor residual registration of 1/8 of the resolution element leads to a standard deviation of around 23° and 42° for SNR = ∞ dB and SNR = 10 dB, respectively.

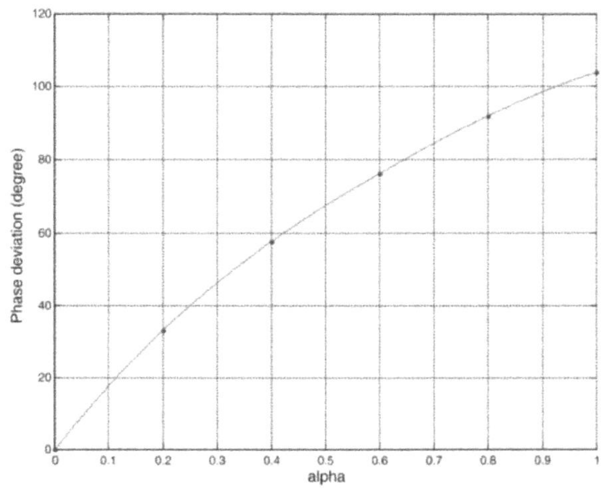

Figure 11.5. *The phase deviation according to the relative change between the two images as a fraction of a resolution element*

The phase deviation depends on the relative change between the two images α of a resolution element:

– if there is a poor geometric registration, then a great relative change involving a large phase deviation, in this case inferometric phase is disturbed;

– the phase deviation σ_ϕ equal to zero for a $\alpha = 0$.

ψ_{20} and ψ_{02}: divergence in the azimuth and in the gate; only the azimuth case is considered.

ψ: phase error at the edges of the bandwidth.

μ: $B_\mu / 2$

$$\psi_{20} = \frac{2}{\pi B_\mu^2} \cdot \psi$$

So, we have:

$$\psi(\mu, v) = 2\pi\psi_{20}\mu^2 = 4\frac{\psi\mu^2}{B^2} \qquad [11.21]$$

For this, the bias phase is obtained:

$$\varphi_0 = \arctan\frac{\xi\left(\sqrt{2\psi/\pi}\right)}{\varsigma\left(\sqrt{2\psi/\pi}\right)} \qquad [11.22]$$

The size of the correlation coefficient

$$|\gamma| = \frac{1}{\sqrt{2\psi/\pi}} \cdot \sqrt{\varsigma^2\left(\sqrt{2\psi/\pi}\right) + \xi^2\left(\sqrt{2\psi/\pi}\right)} \qquad [11.23]$$

where $\varsigma(\ldots)$, $\xi(\ldots)$ represent the Fresnel intervals.

Evidently, the divergence introduces the bias phase.

a)

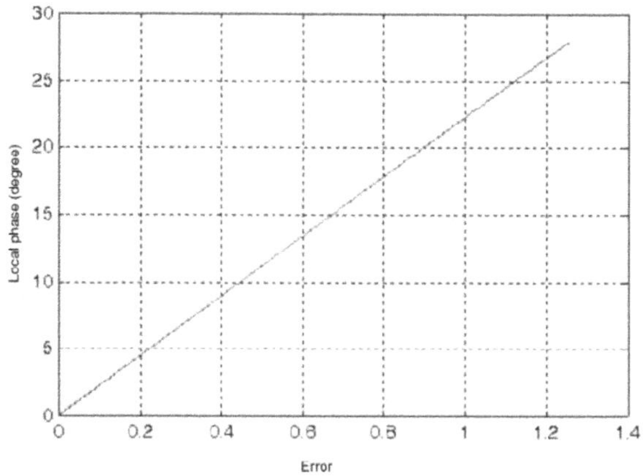

b)

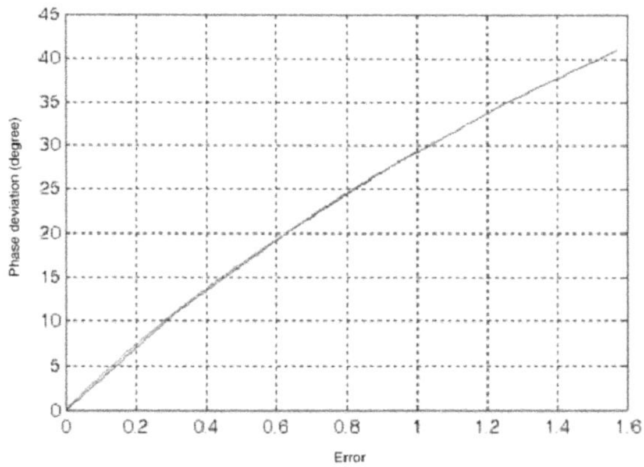

Figure 11.6. *a) The bias phase according to the error phase at the edges of the azimuth bandwidth; b) the phase deviation according to the phase error at the edges of the azimuth bandwidth*

We see on these figures the influence of the phase error at the edges of the azimuth bandwidth on the bias phase and the phase deviation:

– the increase in the bias phase and the phase deviation σ_ϕ has an increase of ψ;

– the deviation $\sigma_\phi = 10°$, the bias phase $\phi_0 = 7°$ for a $\psi = 0.1\ \pi$.

ψ_{11}: Migration of the residual gate in fractions of β of the gate's resolution element $\dfrac{1}{\beta_V}$ and $\psi_{11} = \dfrac{\beta}{\beta_\mu \beta_v}$

with

$$\psi(\mu,\nu)=2\pi\psi_{11}.\mu.\nu=2\pi.\beta.\frac{\mu.\nu}{B_\mu B_\nu} \quad [11.24]$$

we obtain

$$\phi_0=0 \quad [11.25]$$

and

$$|\gamma|=\frac{2}{\pi\beta}\int_0^{\pi\beta/2}\frac{\sin(\mu)}{\mu}d\mu=\frac{Si\left(\pi\beta/2\right)}{\pi\beta/2} \quad [11.26]$$

Si(…) is the integral sinus.

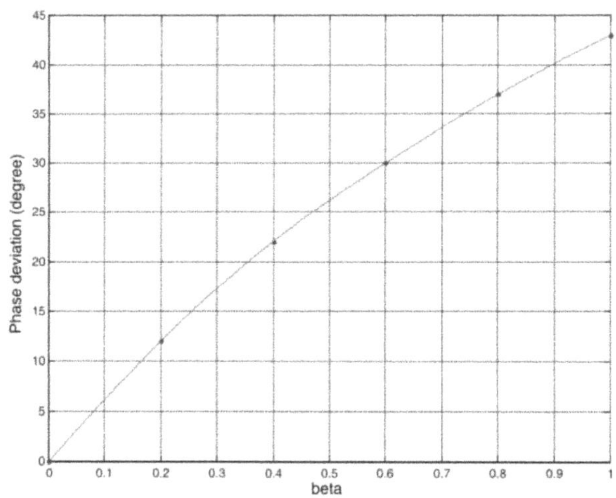

Figure 11.7. *The phase deviation migration of an incorrect resolution element's linear gate*

Figure 11.7 shows the variation in the phase deviation for the incorrect linear gate β's migration between 0 and 1.0 of a resolution element:

– so that the deviation can be null, β should be null;

– we note that at each increase of β, there is an approximate increase of σ_ϕ on the other hand.

ψ_{21}: migration of the non-compensated quadratic gate.

The residual quadratic gate is expressed in fractions ε of a resolution element $1/B_\nu$'s gate.

$$\psi_{21}=\frac{4\varepsilon}{B_{\mu}^2 B_\nu}$$

$$\psi\ (\mu,\nu)=2\,\pi\psi_{21}\mu^2.\nu=\ 8\pi.\varepsilon.\frac{\mu^2.\nu}{B_\mu^2.B_\nu} \qquad [11..27]$$

We simply find that there is no bias phase.

$$\phi_0=0 \qquad [11.28]$$

and

$$|\gamma|=\frac{1}{\sqrt{\varepsilon}}\int_0^{\sqrt{\varepsilon}}\sin c\left(\mu^2\right)d\mu. \qquad [11.29]$$

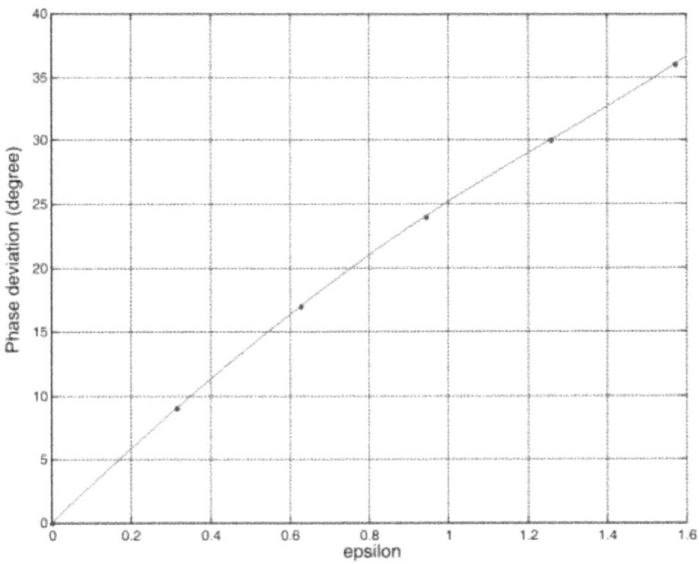

Figure 11.8. *The phase deviation compared to the non-compensated quadratic gate's migration expressed in fractions ε*

Figure 11.8 represents the variation in the phase deviation σ_ϕ with the variation in fractions ε of the resolution element's gate corresponding to the non-compensated quadratic gate's migration:

– if there is no ε, we do not obtain any σ_ϕ deviation;

– σ_ϕ grows with the increase in ε, knowing that we have taken quite small aberration values;

– for ε is equal to 0.5π, σ_ϕ to be augmented to 34°.

11.6. Conclusion

InSAR is a remote sensing technique used for imaging. This radar is made up of two conventional antennae mounted simultaneously on the same platform which transports them along the trajectory. InSAR's functioning is developed on the basis of measuring the phase difference between the two signals received at the two antennae:

– the returns by different positions are recorded and processed to make antennae of the desired length;

– the image phase is sensitive and should be monitored by the imaging process with the aim of reducing the errors produced by the temporal decorrelation phenomenon and the phase aberrations introduced when the data are acquired and processed;

– studying and simulating phase statistics has enabled us to determine the influence of different parameters on the phase deviation, such as the signal to noise ratio and the correlation coefficient;

– we have calculated the inferometric phase's probability density function and typical aberrations in processing, such as poor geometric registration, lack of focalization and the gate's uncompensated migration under the final noise's influence. Only one of these, divergent aberrations, causes a biased phase, but all the others introduce the phase variance which limits the inferometric measures' phase precision;

– these results can form an aide to designing and developing practical criteria likely to respond to demands to model InSAR processes.

12

Telecommunications Networks

12.1. Introduction

Multihoming is a concept that has proved itself in wired networks for transporting information on the Internet, but its integration in mobile networks comes to make the two types of protocol cohabit. In the first place, mobile protocols ensure the information routing function, then the protocol that introduces *Multihoming* [IYE 06, CHA 07].

In this chapter, we will try to integrate the concept of *multihoming* into *ad hoc* mobile networks by using the BGP protocol. First we will use it with the category of reactive protocols such as AODV, then with those of proactive protocols such as OLSR. Finally, our conclusion will be to define the protocol family with which it will perform best.

In order to meet our objective, we will base our study on simulating an *ad hoc* network implemented with the software OPNET. This will enable us to collect a number of statistics illustrated by curves, in order to facilitate their interpretation.

The essential objective of *Multihoming* is to increase the strength and availability of routes for combating the drawbacks of the nodes' mobility, as by definition the *ad hoc* networks are dynamic by nature, which involves a constant change in topology. Consequently, during simulation, the mobility models play a vital role.

12.2. Describing the *ad hoc* simulated network's topology

The software OPNET provides the possibility of realizing several scenarios for a single received topology. Quite evidently, each scenario carries technical modifications. This enables the results to be compared, and the way to be opened for scientific analysis.

Thanks to the software OPNET, we will be able to simulate the performance of an average-sized *ad hoc* network similar to that of a campus. This extends to a surface of 16 km^2.

This *ad hoc* network is made up of nine mobile stations of the *MANET_station_adv* type supporting the protocols IP, IEEE802.11 with the communication interfaces WLAN 1, 2,5.5, 11 Mbps [BRA 01].

Two main elements can be distinguished from this topology: a source node and another destination. Between these two, information messages are transmitted that spread to all the nodes in the network. Each node is characterized by an IP address that will soon be used for simulation purposes.

On the other hand, a mobile network is called *multiresident* when it is simultaneously connected via several MRs; or when an MR has several external interfaces. Therefore, to integrate *multihoming* into this *ad hoc* network, three type *Wlan_ethernet_router* wireless routers have been added to the previous topology, with a WLAN IEEE 802.11 interface. The mobile router is a special node in the *ad hoc* network, as it enables several mobile stations to be connected at the same time. Moreover, it calculates the most optimal routes following the *ad hoc* protocol used, and the network's topology at a given moment.

Additionally, these routers form important elements for simulation, as they are configured so as to support the BGP protocol [GRI 02, NYK 02]. The BGP protocol uses the notion of AS. Therefore, three ASs are defined at each mobile router.

To organize information exchange between the different mobile routers well, a list of neighbors is configured for each (based on the IP addresses) [BET 01, WAT 06].

12.3. The different scenarios enacted

To reveal the importance of *multihoming* integrated into such a mobile topology, our simulation will be organized in two parts. Each part is made up of three scenarios; comparison of results will happen in succession between the first, the second and then the third scenario.

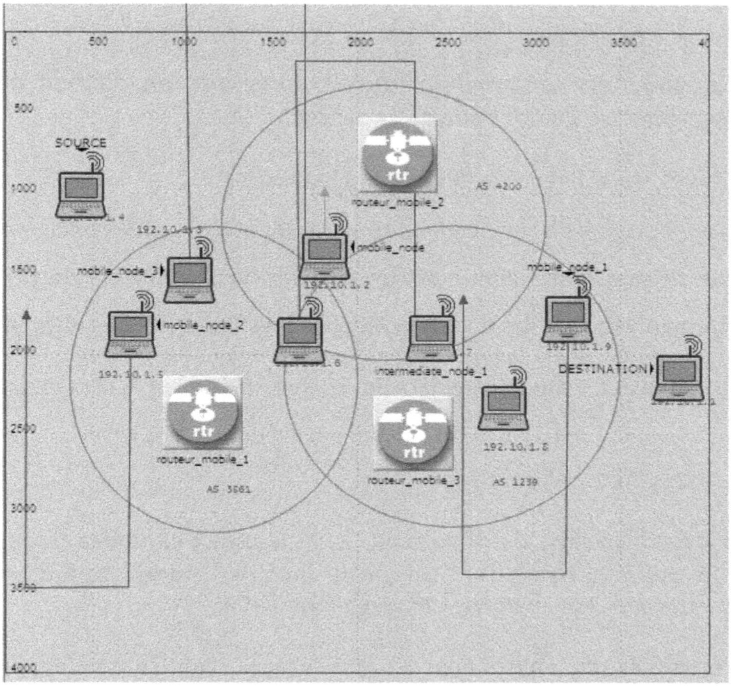

Figure 12.1. *Ad hoc topology network introducing the concept of Mutihoming*

12.3.1. *First part: integrating multihoming in an ad hoc network using the reactive AODV routing protocol*

In this part, we will carry out the following scenarios:

1) *ad hoc* network with simple mobility and without the concept of *multihoming;*

2) *ad hoc* network with simple mobility and with *multihoming* integrated;

3) *ad hoc* network with developed mobility and with *multihoming* integrated.

Mobility models in a simple case are called *VECTOR*. In this case, the mobile station moves upwards in a linear fashion.

For developed mobility, there is the model *MANET_up_right* and MANET *_down _left*. Both show a more complex mobility than *VECTOR*, where the mobile station moves more quickly: either up or down, to left or right, making longer trajectories. This causes a rapid change in the network's topology, indeed even the loss of connection between several stations [KUM 04, COR 99, DHA 05].

12.3.2. *Second part: integrating multihoming into an ad hoc network using the proactive OLSR routing protocol*

For this part, we will recreate the following scenarios:

1) *ad hoc* network with simple mobility and without the *multihoming* concept;

2) *ad hoc* network with simple mobility and with *multihoming* integrated;

3) *ad hoc* network with developed mobility and with *multihoming* integrated.

For the mobility, the models are the same as those described in the first part.

12.4. The statistics collected

Before even launching the simulation, it is necessary to define the group of statistics to collect. These last data will then be stored in a file called *Scalar Statistics* and then interpreted in graph form [GOF 00].

To demonstrate the integration of *Multihoming* and the routing protocols category that would be best for it statistically, we will collect global statistics that include the totality of nodes in the network, as well as the individual statistics for a single node [AHU 00, BOR 06, CAR 03].

12.4.1. *First part*

1) OVER ALL:

– AODV ROUTING TRAFFIC SENT: shows the set of routing packets sent on the entire network for route discovery and maintenance operations such as *RREP, RREQ, RERR* messages. This statistic is measured in packets/second.

2) INDIVIDUAL NODE:

– *Source node IP TRAFFIC SENT*: the set of all the packets sent on the network by this node in packets/second;

– *Destination node IP TRAFFIC RECEIVED*: the set of all the packets received by this node across the network in packets/second;

– *Router mobile 3 nodes IP TRAFFIC RECEIVED:* this statistic is collected for scenarios two and three. It represents the set of all the packets received by mobile router 3 in the topology across the network in packets/second.

NOTE 12.1.–

We have chosen to take only the statistics from the third mobile router (at l' AS 1239) due to its proximity to the destination. This makes this an important node, through which most of the data traffic sent to its destination will pass.

12.4.2. *Second part*

1) GLOBAL

– *OLSR ROUTING TRAFFIC SENT:* shows the set of routing packets sent over the entire network to bring to light paths such as the messages: "*Hello, TC sent, TC forwarded.*" It is measured in packets/second.

2) INDIVIDUAL NODE:

– Source node IP TRAFFIC SENT;

– Destination node IP TRAFFIC RECEIVED;

– Mobile router nodes 3 IP TRAFFIC RECEIVED.

NOTE 12.2.–

To show the difference that exists between the proactive and reactive routing protocols, we should program the start of the packets' transmission by the source after 100 seconds from the start of the simulation, which will last 500 seconds in all. The size of each packet is 1024 bits.

12.5. Discussion of results

12.5.1. *First part: network using AODV for routing*

This first part will be a question of using two protocols that function in conjunction with one another, BGP and AODV. In the case of BGP, it updates the list of three ASs (3561, 4200, 1239) excluding the loops. Communication between these three ASs occurs with the sessions TCP. The BGP is also characterized by the fact that it avoids using periodic update messages only during a change in topology. The BGP sends the updates to the best paths, but in this way it does not have an overall vision of the topology. It is here that the AODV protocol becomes involved to discover and prevent changes in topology. AODV carries out two operations: discovering routes with RREQ, RREP messages and route maintenance with RERR type messages

12.5.1.1 *Comparison between the first and second scenarios*

In Figure 12.2, we can see an absence of traffic during the first 100 seconds due to the absence of packets generated by the source. As we know, AODV is a reactive routing protocol that uses its diffusion mechanism in the network to discover the routes only at the request of the sending nodes.

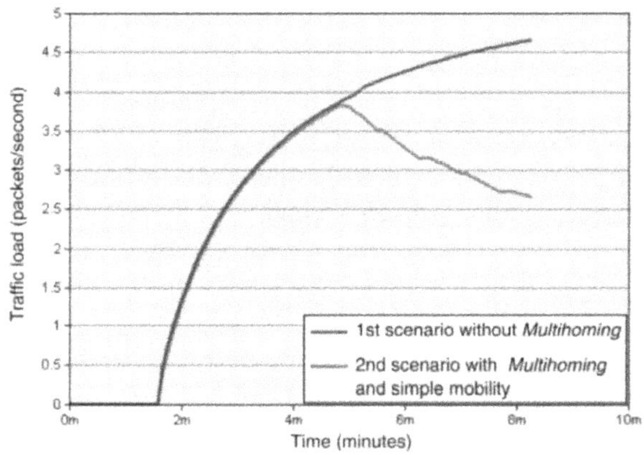

Figure 12.2. *Routing traffic (AODV messages) sent by all the nodes in the network at packets/sec. For a color version of this figure, see www.iste.co.uk./benslama/quantum.zip*

Then, the two curves are almost identical for both scenarios until second 305 of the simulation time.

We can attribute this sudden change to the effects of mobility, as in the first scenario, which does not include the *multihoming* concept, the change in topology will cause the loss of a number of routes, thus causing a significant drop in routing traffic. The latter is due to the method of AODV maintenance, which uses a field called a sequence number at each request; in order to avoid infinitely retransmitting the same request. Therefore if a node receives several copies of the same RREQ, only the first is kept.

In contrast to the second scenario, which integrates the concept of *multihoming*, the information traffic continues to increase (message RREQ, RREP, RRER) notably with mobility, reaching a value of 4.65 packets/sec at the end of the simulation, explained by the availability of links.

Figures 12.3 and 12.4 give an overview of the set of packets that are generated by the source, and those received by the destination node in the two scenarios.

In a general way, these two figures show almost the same evolutions as the previous figure: the absence of traffic for the first 100 seconds, then the curves representing the two scenarios become almost identical, apart from a few small differences, then the sudden drop in traffic in the 1st scenario (not including the *Multihoming*) which starts at the 305th second, whereas for the second it is an increase reaching .55 packets/sec sent by the source, and 1.56 packets/sec received by the destination.

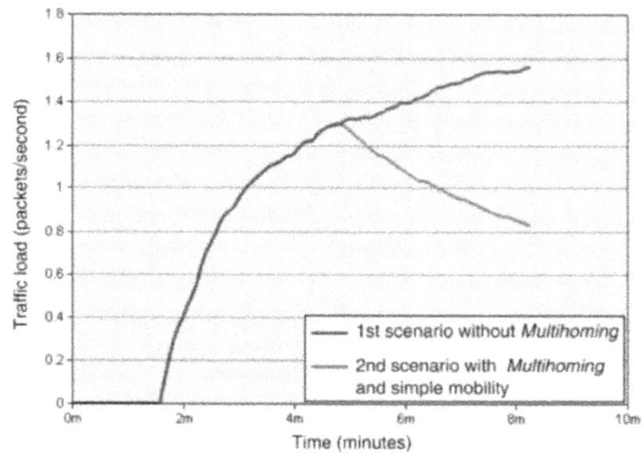

Figure 12.3. *The traffic sent by the source in both scenarios (packets/sec). For a color version of this figure, see www.iste.co.uk./benslama/quantum.zip*

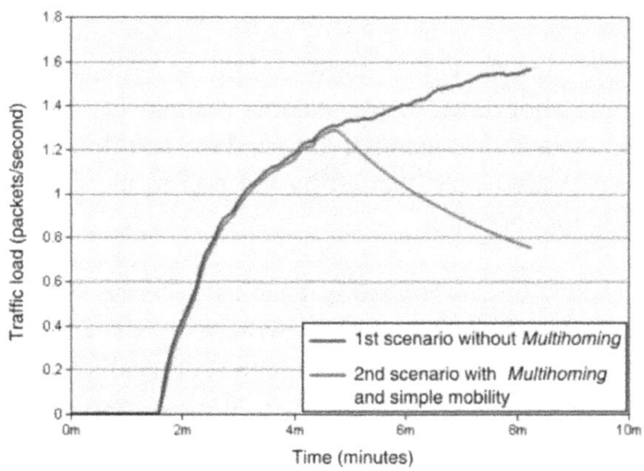

Figure 12.4. *Traffic received by the destination node in both scenarios (packetts/sec). For a color version of this figure, see www.iste.co.uk./benslama/quantum.zip*

12.5.1.2. *Comparison between the second and third scenarios*

Previously, when comparing the 1st and the 2nd scenarios, the importance of *Multihoming* was realized. In order to demonstrate its advantages on an *ad hoc* network that changes rapidly and suddenly in topology, we will compare scenario 2 using a simple *VECTOR* mobility model, and scenario 3 using the developed mobility models: *MANET_up_right*and *MANET _down _left*, as the modification in topology is directly linked to the stations' speed of movement and the network's range of action.

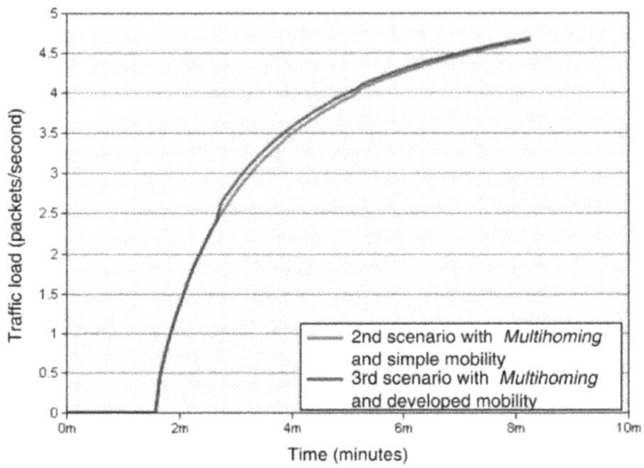

Figure 12.5. *AODV packets sent throughout the network. For a color version of this figure, see www.iste.co.uk./benslama/quantum.zip*

In Figure 12.5, we can see the absence of traffic from the start of the simulation up to the 100th second. Then the curves show that the two scenarios are similar until second 170 of the simulation time; when we note a slight increase in the curve for scenario 3. We can explain this increase by sending additional messages for the route's maintenance and discovery, caused by the nodes' mobility, involving rapid changes and impacting the topology.

Only the integration of *multihoming* enables a node to maintain several communication interfaces at the same time. Where a link is lost, the node tips automatically onto the next, without having to send route maintenance and discovery messages. Avoiding loading the network with useless messages is one of the advantages of the AODV routing protocol.

As for Figures 12.6 and 12.7, they respectively represent the traffic sent by the source and that received by the destination. On the two figures, we can see a perfect similarity between the two curves. These last two demonstrate that the traffic load sent and received has not changed despite the effects of the developed mobility, which cause a rapid change in the topology and several line breaks.

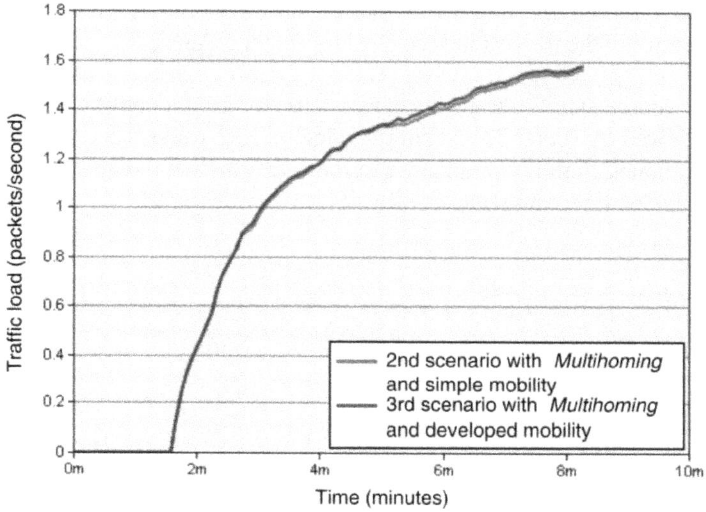

Figure 12.6. *Packets received by destination. For a color version of this figure, see www.iste.co.uk./benslama/quantum.zip*

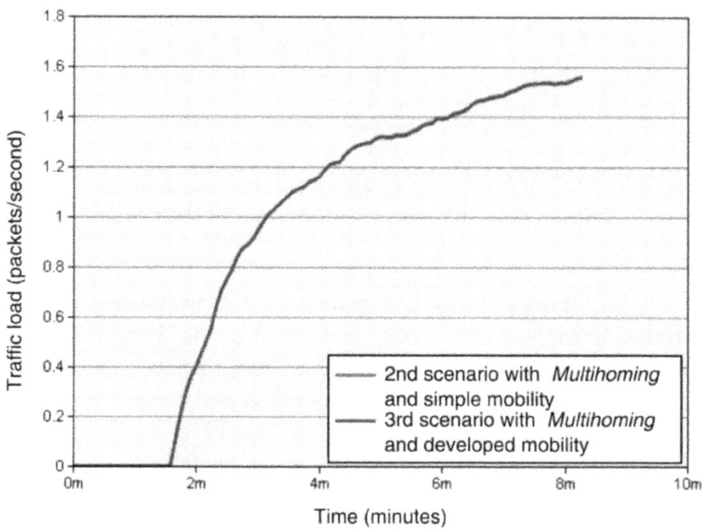

Figure 12.7. *Packets sent by source. For a color version of this figure, see www.iste.co.uk./benslama/quantum.zip*

In fact, the routes between two mobile stations should be constantly adapting to the change in topology to maintain continuity in communication. This involves an increase in the traffic load beside the RREQ and RERR messages. But with the integration of *Multihoming*, this traffic can be shown in Figure 12.8, presenting a comparison between the traffic load received by router 3 AS 1239.

When the source starts to transmit the packets, we note a significant increase in the traffic received by router 3, reaching a value of 1.82 packets/sec in the scenario involving developed mobility, compared to the scenarios with simple mobility with a maximal value of 1.51 packets/sec.

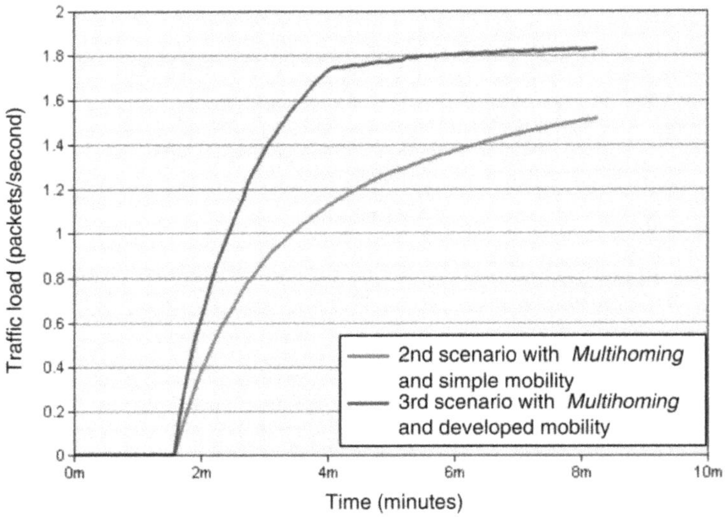

Figure 12.8. *The traffic load received by router 3 in scenarios 2 and 3. For a color version of this figure, see www.iste.co.uk./benslama/quantum.zip*

In order to avoid the problem of congestion, AODV provides a mechanism for repairing a route locally at the node that detects the breakdown, if it has MEMORIZED in its table an alternative route for reaching its destination it uses it and economizes on sending route search and response messages. On the other hand, *multihoming* allows a node to keep several links active at the same time, which enables it to pass automatically from one faulty link to another.

Thus, the significant difference in traffic between the two curves is made up of messages generated by the source transiting via routes to its destination, created by router 3 at AS 1239, and sometimes even in several copies.

This leads us to conclude that integrating *multihoming* has made the *ad hoc* network more robust in the face of the drawback of the nodes' mobility, by increasing route availability.

We can clearly see a sudden drop in traffic in the topologies that do not integrate the concept of *multihoming,* whenever there is a simple mobility. In any case, for networks using the latter concept, the traffic sent and received by the source and destination node remains the same; whether it is simple or developed mobility. The increase in traffic load at the communicating ASs is covered by the means of the BGP protocol.

12.6. Part two: network using OLSR for routing

In this part, we will study the impact of *multihoming* by integrating it on the same topology, except that this time, it uses the proactive OLSR protocol for routing information.

For *multihoming*, the BGP protocol will integrate it with practically the same parameters as the first part. As for the global vision of the network topology, it is the OLSR protocol that is responsible for it.

OLSR uses the concept of MPR multi-point relays, which aims to reduce the number of useless monitoring messages when the network is flooded. This protocol also provides a mechanism for neighbor discovery, achieved by periodically exchanging *HELLO* messages; by contrast, topology messaging is achieved by periodically diffusing *TC sent* and *TC forwarded* messages.

12.6.1. *Comparison between the first and second scenarios*

Observing Figure 12.9, the first observation that comes to mind is automatically and simultaneously sending a large number of routing OLSR packets from the beginning of the simulation, and even before the source has begun to generate packets (after 100th second).

This can be explained by the fact that the OLSR launches a flooding procedure in the entire network to understand its topology, and to enable each node in the network to define its *MPR_SET*.

Moreover, in the same figure we note a large difference between the two curves. As for the first scenario, the traffic load reaches a maximal value of 6.63 packets/sec, but then drops to 5.48 packets/sec.

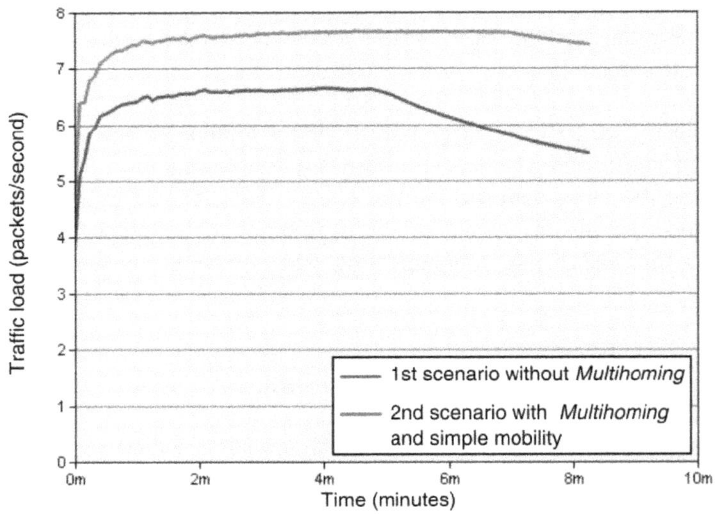

Figure 12.9. *Total traffic routing (OLSR message) on the network. For a color version of this figure, see www.iste.co.uk./benslama/quantum.zip*

For the second curve showing the scenario integrating *multihoming*, the traffic load is significant in comparison to the first curve; as it reaches a maximal value of 7.64 packets/sec, and remains stable for a moment with a slight drop leading to a value of 7.41 packets/sec.

These changes in the two curves are attributed to the fact that in the second scenario, the integration of *multihoming* considerably increases the number of routes available. This involves an increase in the routing messages *HELLO* and *TC*, which are sent periodically to all possible destinations in order to identify the entire topology, and above all the mobility. The nodes should recalculate their own *MPR_SET*.

But for scenario 1, there are fewer routes available, with the mobility causing breaks in the line and so deleting the messages responding to *HELLO* and *TC forwarded.*

Explaining the previous figure enables us to really comprehend Figure 12.10, which shows a comparison between the traffic sent by the source for the same scenarios.

The traffic sent during the first 100 seconds only shows the OLSR routing messages. Then, the increase in traffic is due to the source beginning to generalize packets. But with mobility, an increase in traffic reaches a value of 1.29 packets/sec for the second scenario integrating *multihoming*; whereas for the first, it is a drop to 0.89 packets/sec caused by the breaks in the line between the source and the destination.

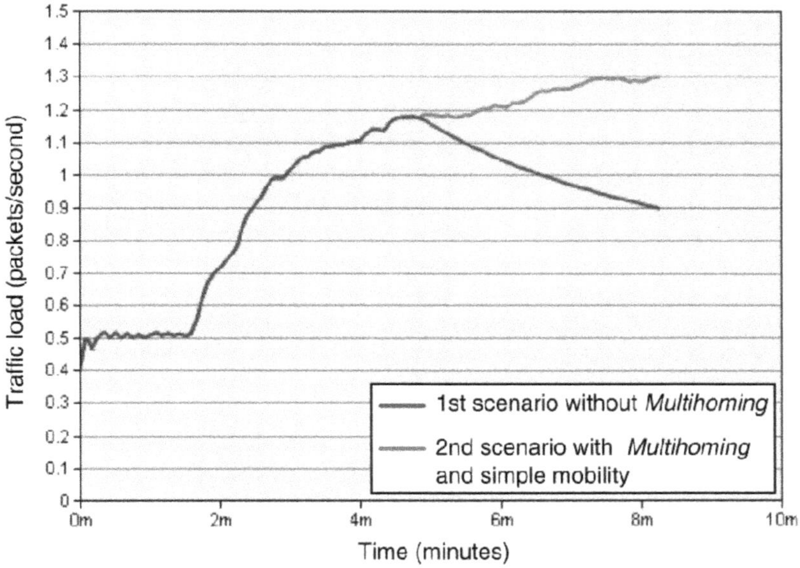

Figure 12.10. *The traffic sent by the source. For a color version of this figure, see www.iste.co.uk./benslama/quantum.zip*

In Figure 12.11, showing a comparison between the traffic received by the source node for the same scenarios; the difference in load is significant as a considerable number of available links triggers an increase in OLSR routing messages for the second scenario whereas for the first, the mobility causes a loss of links and packets sent by the source with a drop in maintenance and neighbor discovery messages.

12.6.2. *Comparison between the second and third scenarios*

Now, we will compare the 2nd and 3rd scenarios to show the effects of developed mobility on such a configuration integrating *multihoming* and using OLSR as a routing protocol.

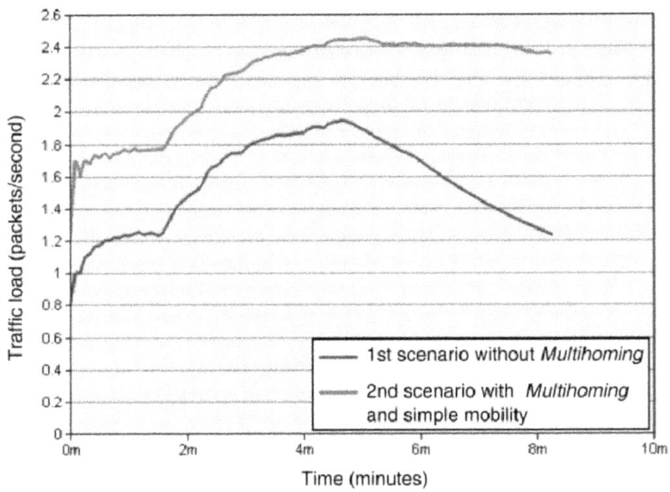

Figure 12.11. *Traffic received by the destination node. For a color version of this figure, see www.iste.co.uk./benslama/quantum.zip*

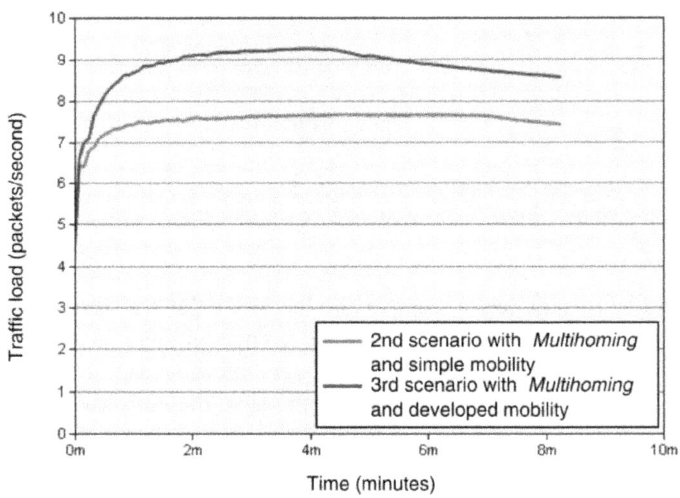

Figure 12.12. *Total routing traffic on the network. For a color version of this figure, see www.iste.co.uk./benslama/quantum.zip*

In Figure 12.12, we see a significant difference between the two curves that show the OLSR routing traffic. Of course, this difference is due to the nodes' mobility, as at each change in topology; the calculations of routes towards all destinations are triggered to reveal the routing tables and recalculate the *MPR_SET*. Topology monitoring and neighbor discovery messages that load the network by the *Multihoming* effect can also be added.

It is important to underline that the slight drop in traffic at the end of the simulation is due to the *MPR_SET*, which only uses symmetrical links; unless the change in topology can make them asymmetrical [PER 03, JAC 01] and [CLA 03].

As for traffic sent by the source, the two curves are perfectly identical for the two scenarios. This demonstrates that simple or developed mobility has no influence on the source where *multihoming* is present.

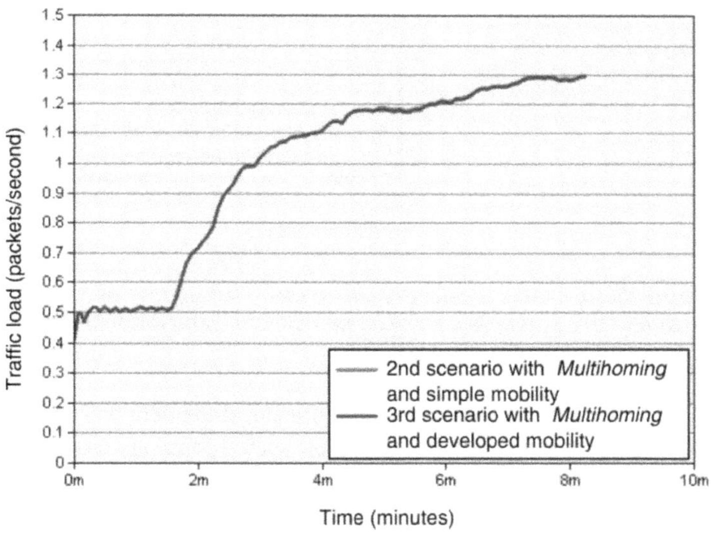

Figure 12.13. *Traffic sent by the source. For a color version of this figure, see www.iste.co.uk./benslama/quantum.zip*

While for the destination, the traffic received (Figure 12.14) in the case of developed mobility is more significant than that of the simple mobility caused by the OLSR routing traffic, as we have explained previously.

In the second part of our simulation, the *multihoming* concept has been integrated into a topology using the proactive OLSR routing protocol.

Generally, proactive routing protocols try to keep the best existing paths to all possible destinations at the level of each node in the network. Since these, combined with a protocol such as BGP, have no global vision of the topology, this will undoubtedly increase the network's strength.

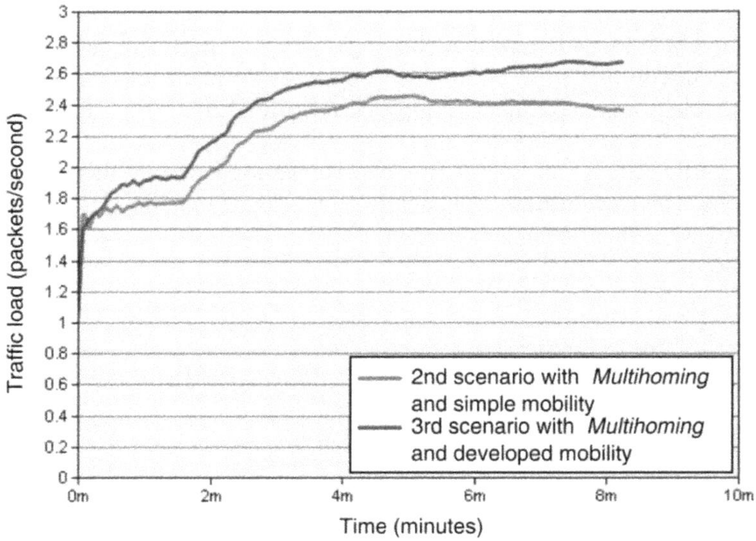

Figure 12.14. *Traffic received by the destination node. For a color version of this figure, see www.iste.co.uk./benslama/quantum.zip*

It is necessary to study the traffic load that transits at the network's different ASs in order to define the impact of the mobility on such a topology. As a sample, we will take the individual statistics from router 3, in AS 1239 due to its proximity to the destination.

In Figure 12.15, we note a significant difference in load between the two curves. For a simple mobility the traffic is relatively low; by contrast for the developed mobility, it is almost double the previous traffic.

We have just attributed this difference in load to the nodes' mobility and the consequent change in topology, which involves recalculating routes towards all the destinations and the *MPR_SET* for each node etc.

But the most important thing is to state that, despite the significant load that transits on AS 1239, the traffic of packets received at the destination node is not the same for both scenarios.

Multihoming enables a node to keep several connections active at the same time. This increases the number of routes available, and with a proactive routing protocol, the routes are safeguarded by a continuous exchange of messages to reveal paths even if they are not used, at the level of each node in the network.

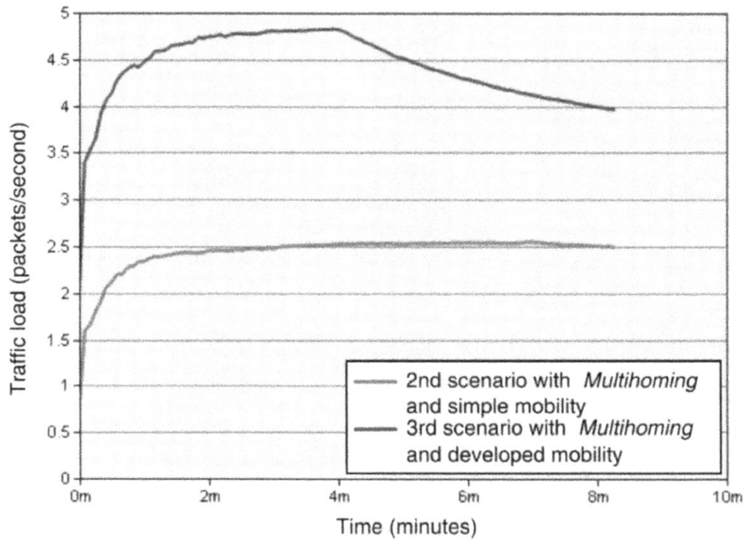

Figure 12.15. *The traffic load received by router 3 in scenarios 2 and 3. For a color version of this figure, see www.iste.co.uk./benslama/quantum.zip*

This leads us to conclude that integrating *multihoming* into an *ad hoc* network using proactive routing protocol makes it stronger by avoiding sudden drops in traffic. Unfortunately, the nodes' mobility influences the traffic received at the destination, characterized by a consequent increase in routing traffic, which is for the most part useless.

12.6.3. *Comparing two parts*

In order to define the *ad hoc* routing protocol category with which *multihoming* will perform better, we have compared the traffic loads received by AS 1239 router number 3 in the first part, which focuses on the reactive protocol category, with that of the second part, on proactive protocols.

In a broader sense, an *ad hoc* routing protocol should find a valid path between the source and the destination for an information packet, taking into consideration the minimization of the network's load and the prevention of concentrations of traffic around certain nodes or links.

In our case, and with the topology that we have defined, the OLSR protocol combined with BGP trigger an increase in the traffic load at AS 1239, and according to Figure 12.16 it will be more than double that generated by the AODV protocol.

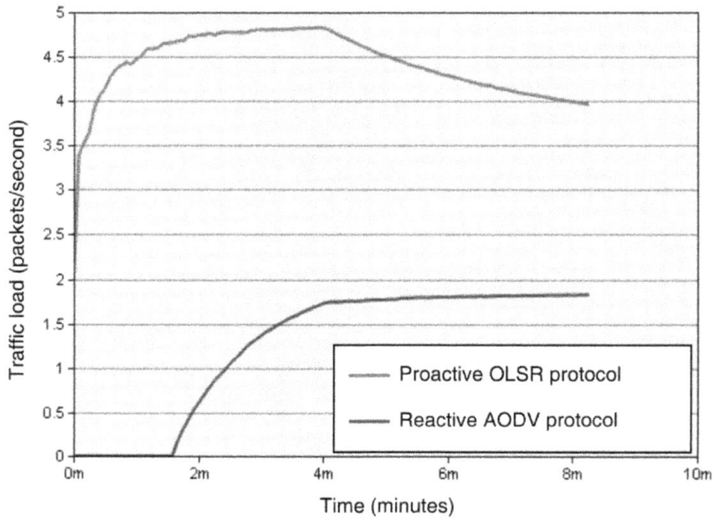

Figure 12.16. *Traffic received by router 3 for both protocols. For a color version of this figure, see www.iste.co.uk./benslama/quantum.zip*

From this figure, it can be concluded that the AODV protocol economizes on transmitting routing messages by keeping the same traffic received by the destination node whatever the extent of the mobility. OLSR, despite the enormous traffic load that transits at the ASs, remains sensitive to frequent changes in topology, which directly influences the traffic received.

This leads us to confirm that integrating *multihoming* with BGP is more advantageous for strength and for load optimization, as well as for resisting the effects of mobility with the reactive AODV protocol and with the proactive OLSR.

12.7. Conclusion

For the *ad hoc* networks, the BGP protocol allows for a node to have a permanent and simultaneous connection to several of its neighbors, and thus to pass automatically from one defective link to another in a good state. Nevertheless, it suffers two major drawbacks:

1) as it no longer has a load balancing system, it does not take account of possible congestion in the routers;

2) it is sensitive to fast oscillation in the routes.

So combining it with a proactive protocol such as OLSR will trigger concentrations of the load at some links, due to the OLSR's topology management mechanism. On the other hand, the cumulative effect of notifications for all the routes available can cause an overload and damage routing stability on the network.

By contrast a reactive routing protocol such as AODV, which provides a mechanism for repairing routes locally at each node, economizes on transmission of route search and response messages; it avoids triggering congestion in the network.

We have finally been able to demonstrate that integrating the *Multihoming* concept into an *ad hoc* mobile network will certainly increase its strength, but it will be more advantageous for the load and for economy of network resources with the reactive routing protocol category than with proactive protocols.

Conclusion

The transition from digital communications to quantum communications, thought in the preamble to be a necessity for responding to expectations of quantum computing, is beginning to take shape though developments that are as academic as they are technological. In this book, we took constant care to reconcile theoretical advances with experimental discoveries. Of course, it was difficult to reconcile two designs that are quite different in their approaches. We have, however, made an effort at synergy to bring the two visions closer. In the literature review, we used the most essential and cited publications. The basic framework of these transitions is signal processing. We had thought that the reality was less linear, and so we sought the most general formulations; this is apparent in the tensor formulation of the digital signal by tensors, and the quantum signal by spinors. Geometrically, the most general formulation is Riemannian geometry and its corollaries (varieties). The results obtained in factorizing numbers caused us to introduce Gauss sums and Clifford algebra.

Of course, many papers were omitted due to their complexity and the non-obvious conclusions, in the practical sense of the term, obtained. With the aim of highlighting this work, we thought it useful to add practical applications in mobile communications in order to highlight the signal's characteristics (variance, covariance, probability, etc.). The study of these characteristics has been extended to STAP and InSAR radar, which still have an obvious attraction for specialists.

One important element has not been sufficiently highlighted in this work: the domain of wavelets and fractals. Through an application on interferences, we have tried to make up for this imbalance. The Dempster–Shafer theory, or evidence theory, has been examined in the context of radar target tracking. Finally, to answer expectations in the domain of networks, we have included *multihoming* and *ad hoc* networks.

The combination of MIMO and OFDM has been designed to improve service quality (QoS) and/or the system's data rate by taking advantage of the multiplexing gain and/or diversity gain. The evolution of wireless standards, such as mobile WiMax (IEEE 802.16e) and WiFi (norm IEEE 802.11n), will use several transmission and reception antennae with an OFDM modulation for increased spectral effectiveness and improved performances. However, the system's performance relies on CSI knowledge at the receiver. Perfect estimations of the channel can only be obtained if the channel is not corrupted by noise and is invariant in time. ICI (inter–carrier–interference) makes perfect estimation of the channel impossible in mobile channels because of its temporal variation and fast channel variations.

We demonstrated the Dempster–Shafer theory, or evidence theory, using probabilistic mathematics, with a formal framework for reasoning with uncertainty, a model that enables knowledge to be modeled. Indeed, via mass functions, which are tools for measuring subjective probability, it is possible to evaluate the degree of truth in an expert opinion. With the introduction of evidence masses, the rule of combination enables information to be processed to make it reliable. This greatly helps decision-making based on maximum credibility.

We have seen performances for multi-target tracking and the association of our method; the results obtained by simulation have demonstrated a good performance for the Dempster–Shafer theory.

We have tackled the problem of designing an approach capable of minimizing interferences, from multiple accesses or from the channel, while still ensuring a good synchronization and a totally secure transmission, then implanting this approach in software under a *C++ builder* environment. This second sequence's role is to vary the spreading sequence's initial condition which is that, for each transmission established, the spreading sequence's initial condition takes the value of this second sequence, for the 1st transmission, it takes the 1st value, for the 2nd transmission, it takes the 2nd and so forth. For the channel's interferences, we had to resort to convolutional coding with Viterbi decoding as well as the discreet wavelet transform as an input filter for the receiver to de-noise the signal received of noise from the channel.

The interferometric radar with synthetic aperture is a remote sensing technology used for imaging. This radar is made up of two conventional antennae mounted simultaneously on the same platform that transports them along the trajectory. The InSAR's functioning is detailed based on measuring the phase difference between the two signals received at the two antennae. The results can form an aide for designing and developing practical criteria likely to answer demands for modeling the InSAR process.

In this application, we have integrated the concept of *multihoming* into mobile networks, in order to increase: their strength, their resistance to breakdowns, and in order to enable load sharing, which has the consequence of increasing flow. To do this, we have chosen the BGP protocol, which will take responsibility for introducing the *multihoming* concept on *ad hoc* MANET networks. The will cause BGP to cohabit with other *ad hoc* routing protocols that are classed in three categories: reactive, proactive and hybrids.

Bibliography

[AI 15] AI X.F., LUO Y.J., ZHAO Q., "Transient interference excision and signal recovery in OTHR using robust principal component", *Electronics Letters*, vol. 51, no. 17, pp.1363–1364, 2015.

[AKY 11] AKYILDIZ I.F., JORNET J.M., PIEROBON M., "Nanonetworks, a new frontier in communications?", *Communications of the ACM*, vol. 54, no. 11, pp. 84–89, 2011.

[ALM 07] ALMEIDA L.F DE, "PARAFAC based unified tensor modeling for wireless communications systems with application to blind multiuser equalization", *Signal Processing*, vol. 87, no. 2, pp. 337–351, 2007.

[ANG 07] ANGELAKIS D.G., SANTOS M.F., BOSE S., "Photonic phase transistions, spin models, and QIP in coupled cavity arrays", *IEEE International Quantum Electronics Conference*, CLEOE-IQEC, p.1, 2007.

[ANT 12] ANTONOSYAN D.A., TAMAZYAN A.R., KRYUCHKYAN G.Y., "Chirped quasi phase matching with Gauss sums for production of biphotons", *J. Phys. B: At. Mol.Opt.Phys.*, vol. 45, pp. 1–7, 2012.

[ARC 05] ARCE G.R., *Nonlinear Signal Processing, a Statistical Approach*, Wiley Interscience, 2005.

[ARM 02] ARMENGAUD A.A., GUYON G., CAMPISTRON B., Analyse opérationnelle des données du radar UHF de Marignane, available at: www.airpaca.org, 2002.

[APP 91] APPRIOU, A., "Probabilités et incertitude en fusion de données multi-senseurs", *Revue Scientifique et Technique de la Défense*, vol. 11, pp. 27–40, 1991.

[ATH 94] ATHAS W.C., SVENSON L.J., KOLLER J.G. *et al.*, "Low power digital systems based on adiabatic switching principles", *IEEE Transactions on Very large Scale Integration (VLSI) Systems*, vol. 2, no. 4, pp. 398–407, 1994.

[BAD 05] BADEAU R., DAVID B., RICHARD G., "Fast approximated power iteration subspace tracking", *IEEE Transactions on Signal Processing*, vol. 53, pp. 2931–2941, August 2005.

[BAL 97] BALDWIN J.F., LAWRY J., MARTIN T.P., "A mass assignment based ID3 algorithm for decision tree induction", *Int. J, Intell. Systems*, vol. 12, pp. 523–552, 1997.

[BAM 98] BAMLER R., HARTL P., "Synthetic aperture radar interferometry", *Inverse problems*, vol. 14, pp. R1–54, 1998.

[BAN 12] BANERJEE S., BHATTACHARYA B.B., DAS S. *et al.*, "On the construction of a generalized Voronoi inverse of a rectangular tessalation", *9th Intenational Symposium on Voronoi Diagramms in Science and Engineering*, pp. 132–137, 2012.

[BAR 10] BARACHANT A., BONNET S., CONGEDO M. *et al.*, "Common spatial pattern revisited by Riemannian geometry", *IEEE International Workshop on Multimedia Signal Processing (MMSP)*, pp. 472–476, 2010.

[BAR 88] BAR-SHALOM Y., FORTMANN T.E., *Tracking and Data Association*, Academic Press, 1988.

[BAR 91] BARKAT M., VARSHNEY P.K., "Adaptive Cell-Averaging CFAR detection in distributed sensor networks", *IEEE Transactions on Aerospace and Electronic systems*, vol. 27, no. 3, pp. 424–429, 1991.

[BAS 08] BASSINGTHWAIGHTE J.B., CHIZECK H.J., "The Physiome Projects and multiscale Modeling", *IEEE Signal Processing magazine*, pp. 121–124, 2008.

[BAU 01] BAUGH J., FINKELSTEIN D.R., GALIAUTDINOV A. *et al.*, "Clifford algebra as quantum language", *Journal of Mathematical Physics*, vol. 42, pp. 1489–1500, 2001.

[BAU 96] BAUTISTA R., CRISCUOLO A., DURDEVIC M. *et al.*, "Quantum clifford algebras from spinors representations", *Journal of Mathematical Physics*, vol. 37, pp. 5747–5775, 1996.

[BEH 94] BEHRENS R.T., SCHARF L.L., "Signal processing applications of oblique projections operators", *IEEE Transactions on Signal Processing*, vol. 42, no. 6, pp. 1413–1424, 1994.

[BEN 00] BENNET C., DIVINCENZO D.P., "Quantum information and computation", *Nature*, vol. 404, pp. 247–255, 2000.

[BEN 15a] BENSLAMA M., BATATIA H., BOUCENNA M.-L., *Ad Hoc Networks Telecommunications and Game Theory*, ISTE, London and John Wiley and Sons, New York, 2015.

[BEN 15b] BENSLAMA M., BATATIA H., KIAMOUCHE W., *Connections Management Strategies in Satellite Cellular Networks*, ISTE, London and John Wiley and Sons, New York, 2015.

[BER 98] BERNDT B.C. *et al.*, *Gauss and Jacobi Sums*, Wiley Interscience Publications, vol. 21, 1998.

[BER 99] BERGER S.D., WELSH B.M., "Selecting a reduced-rank transformation for STAP, a direct form perspective", *IEEE Transactions on Aerospace and Electronics Systems*, vol. 35, no. 2, pp. 722–729, 1999.

[BET 01] BETTSTETTER C., "Mobility modeling in wireless networks : categorization, smooth movement, and border effects", *ACM SIGMOBILE Mobile Computing and Communications Review*, 2001.

[BET 03] BETH J., LEUCHS G., *Quantum Information Processing*, Wiley, 2003.

[BIH 11a] BIHAN LE N., Contributions au traitement des signaux à valeurs sur des structures algébriques non commutatives, Thesis, University of Grenoble, 2011.

[BIH 11b] BIHAN LE N., SANGWINE S.J., "Analyse de signaux vectoriels basée sur le modèle quaternionique", *19ème Colloque sur le Traitement du Signal et des Images*, pp. 304–307, 2003.

[BIH 11c] BIHAN LE N., AMBLARD P.O., "Detection and estimation of gaussian proper quaternion valued random processes", available at: ima.org.uk/_dbl–documents/Le_Bilan_2.pdf, 2011.

[BIN 03] BINZ E., PODS S., SCHEMPP W., "Heisenberg groups, a unifying structure of signal theory, holography, and quantum information theory", *Journal of Applied Mathematics and Computing*, vol. 11, no. 1, pp. 1–57, 2003.

[BHA 14] BHASKAR V., "Novel closed form expressions for ergodic capacity of MIMO-OSFBC-OFDM systems over channel adaptation policies", *Wireless Pers. Commun*, vol. 77, pp. 811–822, 2014.

[BOC 99] BOCHE H., "On the calculation of the Hilbert Transform", *IEEE International Symposium on Circuits and Systems*, vol. 3, pp. 183–186, 1999.

[BOG 10] BOGUNOVIC N., JOVIC A., "Processing and analysis of biomedical nonlinear signals by data mining methods", *17th International Conference on Systems, Signals and Image Processing*, p. 276, 2010.

[BOR 06] BORGIA E., CONTI M., DELMASTRO F., "Mobile MAN; Design, Integration and Experimentation of Cross-layer Mobile Multi-hop Ad Hoc Networks", *IEEE Communication Magazine, Ad Hoc sensor Network Series*, vol. 44, no. 7, 2006.

[BRA 01] BRASWELL B.E., MCEACHEN J.C., "Modelling data rate agility in the IEEE 802.11a WLAN Protocol", *OPNETWORK* March 2001.

[BRA 06] BRACKX F., SCHEPPER De N., SOMMER F., "The two dimensional Clifford–Fourier transform", *Journal of Mathematical Imaging and Vision*, vol. 26, no. 1, pp. 5–18, 2006.

[BRA 79] BRASSEL K.E., REIF D., "A procedure to generate Thiessen polygons", *Geographical Analysis*, vol. 11, no. 3, pp. 289–303, 1979.

[BRI 59] BRILLOUIN L., *Science and Information Theory*, 2nd ed., Academic Press, New York, 1962.

[BRU 01] BRUKNER C., ZEILINGER A., "Conceptual inadequacy of the Shannon information in quantum measurements", *Phys. Rev. A*, vol. 63, no. 022113, 2001.

[BRU 14a] BRUNO N., CRUZERO E.Z., MARTIN A. *et al.*, "Simple pulsed polarization entangled photon pair source", *Optics Communications*, vol. 327, pp. 3–6, 2014.

[BRU 14b] BRUNO N., MARTIN A., THEW R.T., "Generation of tunable wavelength coherent states and heralded single photons for quantum optics applications", *Optics Communications*, vol. 327, pp. 17–21, 2014.

[BUR 06] BURKARDT J., GUNZBURGER M., LEE H.C., "Centroidal Voronoi tessellation based reduced order modeling of complex systems", *SIAm, J.Sci. Comput.*, vol. 82, no. 2, pp. 459–484, 2006.

[CAR 90] CARDOSO J.F., "Eigen structure of the fourth order cumulant tensor with application to the blind source separation problem", available at: people.ee.duke.edu/~/carin/cardoso_ICASSP90.pdf, 1990.

[CAR 91] CARDOSO J.F. "Super symmetric decomposition of the fourth order cumulant tensor blind identification of more sources than sensors", available at: faculty.cua.edu/regalia/EE717/cardoso.pdf, 1991.

[CAS 07] CASTANO–MORAGA C.A., "A Riemannian approach to anisotropic filtering of tensors fields", *Signal Processing*, vol. 87, no. 2, pp. 263–276, 2007.

[CIC 08] CICHOCKI A., ZDUNEK R., AMARI S.I., "Nonnegative matrix and tensor factorization", *IEEE Signal Processing Magazine*, vol. 25, no. 1, pp. 142–145, 2008.

[CIC 14a] CICHOCKI A., "Era of big data processing, a new approach via tensor networks and tensor decomposition, available at: arxiv.org/pdf/1403.2048.pdf, 2014.

[CIC 14b] CICHOCKI A., MANDIC D., LATHAUWER DE L., "Brain decompositions for signal processing applications. From two way to multi way component analysis", *IEEE Signal Processing Magazine*, vol. 32, no. 2, pp. 145–163.

[CHA 05] CHANG C.J., HUANG C.W., LIN Y.P. *et al.*, "FPGA platform for CPU design and applications", *Proceedings of 5th IEEE Conference on Nanotechnology*, Nagoya Japan, pp. 1–4, 2005.

[CHA 05a] CHANG F.S., CHEN H.T., CHAO K.C. *et al.*, "A novel monopole antenna for WLAN access point applications", *International Symposium on Antenna and Propagation*, vol. 2A, pp. 438–441, 2005.

[CHA 05b] CHANG F.S., CHEN H.T., HSU P.H., "A novel broadband antenna for WLAN applications", *International Symposium on Antenna and Propagation*, vol. 2A, pp. 557–560, 2005.

[CHA 07] CHAROENPANYASAK S., PAILLASSA B., "SCTP multihoming with Cross Layer Interface in ad hoc Multihomed Networks", *IEEE International Conference on Wireless and Mobile Computing, Networking and Communications (WiMob)*, New York, USA, 2007.

[CHE 07] CHEN J., WONG S.T.C., "Nanotechnology for genomic signal processing in cancer research", *IEEE Signal Processing Magazine*, pp. 111–121, 2007.

[CHO 04] CHOU E.Y., HUANG J.C., HUNG M.S. *et al.*, "Baud rate channel equalization in nanometer technologies", *IEEE Transactions on Very Large Scale Integration (VLSI) Systems*, vol. 12, no. 11, pp. 1174–1181, 2004.

[CHU 94] CHUNG J., CHUNG S.Y., KIM D., "A characterization for Fourier hyperfunctions", *Publ. RIMS Kyoto University*, vol. 30, pp. 203–208, 1994.

[CHU 95a] CHUNG S.Y., KIM D., "Distributions with exponential growth and Bochner–Schwartz theorem of Fourier hyperfunctions", *Publ. RIMS Kyoto University*, vol. 31, pp. 829–845, 1995.

[CHU 95b] CHUNG J., CHUNG S.Y., KIM D., "Positive definite hyperfunctions", *Nagoya Math. J.*, vol. 140, pp. 139–149, 1995.

[CLI 78] CLIFFORD, "Applications of Grassmann's extensive algebra", *American Journal of Mathematics*, vol. 1, no. 4, pp. 350–358, 1878.

[COO 65] COOLEY J.W., TUKEY J.W., "An algorithm for the machine calculation of complex Fourier series", *Mathematics of Computation*, vol. 19, no. 90, pp. 297–301, 1965.

[COO 08] COOLIDGE J.L., *The Elements of Non-Euclidean Geometry*, Oxford Clarendon Press, 2008.

[CON 97] CONNES A., "Brisure de symétrie spontanée et géométrie du point de vue spectral", *Journal of Geometry and Physics*, vol. 23, no. 3–4, pp. 206–234, 1997.

[COQ 01] COQUEREAUX R., "Classical and quantum polyhedra, a fusion graph algebra point of view", *AIP conference Proceedings*, vol. 581, pp. 181–203, 2001.

[COR 99] CORSON S., MACKER J., "Mobile Ad Hoc Networking (MANET): Routing Protocol Performance Issues and Evaluation Considerations", *RFC 2501*, 1999.

[DEL 34] DELAUNAY B., "Sur la sphère vide", *Bulletin de l'académie des Sciences de L'URSS*, no. 6, pp. 793–800, 1934.

[DEM 67] DEMPSTER A., "Upper and lower probabilities induced by multivalued mapping", *Annals of Mathematical Statistics*, vol. AMS-38, pp. 325–339, 1967.

[DEM 68] DEMPSTER A., "A generalization of bayesian inference", *Journal of the Royal Statistical Society*, Series B, vol. 30, pp. 205–245, 1968.

[DEU 04] DEUBEL M., FREYMANN G.V., WEGENER M. *et al.*, "Direct laser writing of three dimensional photonic crystal templates for telecommunications", *Nature Materials*, vol. 3, no. 44, pp. 444–447, 2004.

[DIM 13] DIMITRIADOU E., ZOIROS K.E., "Proposal for ultrafast all optical XNOR gate using single quantum dot semi conductor optical amplifier based Mach Zender interferometer", *Optics & Laser Technology*, vol. 45, pp. 79–888, 2013.

[DJI 08] DJIKNAVORIAN P., "Fusion d'informations dans un cadre de raisonnement de Dezert–Smarandache appliquée sur des rapports de capteurs ESM", *Mémoire de maitrise*, Université Laval, QUEBEC, 2008.

[DON 05] DONG Y., "Approximate invariance of inverse of covariance matrix and resultant pre built space time non adaptive processing", *Proceeding of the 8th International symposium in signal processing and its applications*, vol. 2, pp. 519–522, 2005.

[DOR 02] DORST L., MANN S., "Geometric algebra, a computational framework for geometrical applications", *IEEE Computer Graphics and Applications*, vol. 22, no. 4, pp. 58–67, 2002.

[DOU 00] DOUGHERTY E.R., KIM S., CHEN Y., "Coefficient of determination in nonlinear signal processing", *Signal Processing*, vol. 80, pp. 2219–2235, 2000.

[DU 15] DU L., WANG B., WANG P. *et al.*, "Noise reduction method based on principal component analysis with beta process for micro Doppler radar signatures", *IEEE Journal of Selected Topics in Applied Earth Observations and Remote Sensing*, vol. 8, no. 8, p.4028, 2015.

[DON 95] DONOHO D.L., "Denoising via soft thresholding", *IEEE Transactions on Information Theory*, vol. 41, pp. 613–627, May 1995.

[EBE 95] EBERHART R., "A new optimizer using particle swarm theory", *6th International Symposium on Micro Machine and Human Science*, pp. 39–43, 1995.

[EFR 82] EFRON B., "The jackknife, the boostrap and other resampling plans", *CBMS-NSF Regional Conference Series in Applied Mathematics*, 1982.

[ELD 01] ELDAR Y.C., Quantum Signal Processing, PhD Thesis, MIT, 2001.

[ELD 02] ELDAR Y.C., "Quantum Signal processing", *IEEE Signal Processing Magazine*, vol. 19, no. 6, pp. 12–32, 2002.

[ELD 09] ELDAR Y.C., "Uncertainty relations for shift invariant analog signals", *IEEE Transactions on Information Theory*, vol. 55, no. 12, pp. 5742–5757, 2009.

[ELI 55] ELIAS P., "Coding for noisy channels", *IRE Conv. Rec*, vol. 3, no. 4, pp. 37–46, 1955.

[ELT 05] ELTOFT T., "The Rician inverse Gaussian distribution, a new model for non Rayleigh signal amplitude statistics", *IEEE Transactions on Image Processing*, vol. 14, no. 11, pp. 722–1735, 2005.

[EST 86] ESTENES D., "A unified language for Mathematics and physics", *Clifford Algebras and their Applications in Mathematical Physics*, pp. 1–23, 1986.

[FAR 02] FARINA A., LOMBARDO P., MARSELLA M., "Joint tracking and identification algorithms for multisensory data", *IEEE Proc. Radar Sonar Navig.*, vol. 149, no. 6, pp. 271–280, 2002.

[FAU 93] FAUGERAS O., *Three Dimensional Computer Vision, A geometric view point*, MIT Press Cambridge, 1993.

[FLE 07] FLETCHER P.T., "Riemannian geometry for the statistical analysis of diffusion tensor data", *Signal Processing*, vol. 87, no. 2, pp. 250–262, 2007.

[FLI 07] FLIESS M., "Critique du rapport signal _a bruit en théorie de l'information || A critical appraisal of the signal to noise ratio in information theory HAL Id: inria-00195987", available at: https://hal.inria.fr/inria-00195987, 2007.

[FON 06] FONTIJNE D., BOUMA T., DORST L., "Gaigen, a geometric algebra implementation generator", *Proceedings of the 45th International Conference on Generative Programming and Component Enginering*, ACM, New York, pp. 141–150, 2006.

[FOR 11] FORNASSIER M., RAUHUT R., "Compressive sensing", *Handbook of Mathematical Methods in Imaging*, pp. 187–228, Springer, 2011.

[FOR 73] FORNEY G.D., "The Viterbi algorithm", *Proceedings of the IEEE*, vol. 61, pp. 268–278, 1973.

[FOR 87] FORTUNE S., "A sweepline algorithm for Voronoi Diagram", *Algorithmica*, vol. 2, no. 1, pp. 153–174, 1987.

[FOU 27] FOURIER J., "Mémoire sur la distinction des racines imaginaires, et sur l'application des théorèmes d'analyse algébrique aux équations transcendantes qui dépendent de la théorie de la chaleur", *Mémoires de l'Académie royale des sciences de l'Institut de France*, vol. 7, pp. 605-624, 1827.

[FRA 11] FRASCA M., LIBERATI R., "Riemann manifolds from Hellinger distance", *IEEE Conference on Advances in Radar and Remote Sensing (TyWRRS)*, pp. 59–61, 2012.

[FRA 13] FRANCHINI S., GENTILE A., SORBELLO F. *et al.*, "Design and implementation of an embedded coprocessor with native support for 5D quadruple based Clifford algebra", *IEEE transactions on Computers*, vol. 62, no. 12, pp. 2366–2381, 2013.

[FRI 14] FRITZCHE S., "The Feynman tools for quantum information processing, design and implementation", *Computer Physics Communications*, vol. 185, pp. 1697–1718, 2014.

[GAB 46] GABOR D., "Theory of communication, Part 1 : the analysis of information", *Journal of the Institute of Electrical engineering, Part III radio and communication Engineering*, vol. 93, no. 26, pp. 429–441, 1946.

[GAB 88] GABRIEL A.K., GOLDESTEIN R.M., "Crossed orbit interferometric: Theory and experimental results from SIR-B", *Int. J. Remote Sensing*, vol. 9, no. 5, pp. 857–872, 1988.

[GAL 13] GALLIER J.H., "Clifford Algebras, Clifford Groups, and a generalization of the quaternions, the Pin and Spin groups", *Technical Reports (CIS)*, University of Pennsylvania, pp. 1–46, 2013.

[GAL 97] GALY J., ADNET C., CHAUMETTE E., "Narrow band source separation in wide band context applications to array signal processing", *IEEE Conference*, pp. 373–376, 1997.

[GAN 05] GUAN X., XIAO Y.I., YOU H.E., "An Improved Dempster-Shafer Algorithm for Resolving the Conflicting Evidences Research Institute of Information Fusion", *China International Journal of Information Technology*, vol. 11, no. 12, 2005.

[GAO 12] GAO W., "Forecasting of nonlinear signal processing", *International Conference on Convergence Information Technology*, vol. 19, pp. 97, 2012.

[GAO 14] GAO J., LIU H., "Dirichlet characters, Gauss sums and arithmetic Fourier transforms", *Appl. Math. J. Chinese Univ.*, vol. 29, no. 3, pp. 307–316, 2014.

[GAO 15] GAO G., WANG X., LAI T., "Detection of moving ships based on a combination of magnitude and phase in Along track interferometric SAR. Part I: SIMP metrics and its performance", *IEEE Transactions on Geoscience and Remote Sensing*, vol. 53, no. 7, pp. 3565–3591, 2015.

[GAU 66] GAUSS C.F., *Disquisitiones Arithmetica Leipzig 181*, Yale University Press, 1966.

[GER 03] GERRISH A., SCHULTEISS P.M., "Information rates of non Gaussian processes", *IEEE Transactions on Information Theory*, vol. 10, no. 4, pp. 265–271, 2003.

[GER 05] GERSHMAN A.B., SIDIROPOULOS N.D., *Space Time Processing for MIMO Communications*, John Wiley and Sons, 2005.

[GER 83] GERTH D.H.G. F II, VAALER J.D., "Gauss Sums and Fourier Analysis on multiplicative subgroups of Zq", *Trans. Am. Math. Soc.*, vol. 227, pp. 711–726, 1983.

[GIA 87] GIANNAKIS G.B., "Cumulants, a powerful tool in signal processing", *Proceedings of the IEEE*, vol. 75, no. 9, pp.1333–1334,1987.

[GIL 08] GILOWSKI M., WENDRICH T., MULLER T. *et al.*, "Gauss sum factorization with cold atoms", *Physical Review letters*, vol. 100, pp. 030201–1,030201, 2008.

[GHE 10] GHEORGHIU V., LOOI S.Y., "Construction of equally entangled bases in arbitrary dimensions via quadratic Gauss sums and graph states", *Physical Review A,* vol. 81, pp. 062341–1,062341, 2010.

[GHO 94] GHOSH S., MALLET R.L., "Voronoi cell finite elements", *Computers & Structures*, vol. 50, no. 1, pp. 33–46, 1994.

[GRE 00] GREEN H.S., *Information Theory and Quantum Physics*, Springer, Berlin, 2007.

[GUP 05] GUPTA P., JHA N.K., "An algorithm for nano-pipeling of RTD Based circuits and architectures", *IEEE Transactions on Nanotechnology,* vol. 4, no. 2, pp. 159–167, 2005.

[GRE 78] GREEN P.J., SIBSON R., "Computing Dirichlet tessellation in the plane", *The Computer Journal*, vol. 21, no. 2, pp. 168–173, 1978.

[GRI 02] GRIFFIN T.G., WILFONG G., "Analysis of the med oscillation problem in BGP", *Proc. of the 10th IEEE Int. Conf. on Network Protocols*, pp. 90–99, 2002.

[GUE 00] GUERCI J.R., GOLDSTEIN J.S., REED I.S., "Optimal and adaptive reduced-rank STAP", *IEEE Transactions on Aerospace and Electronics Systems,* vol. 36, no. 2, pp. 647–663, 2000.

[GUM 35] GUMBEL E.J., "Les valeurs extrêmes des distributions statistiques", *Annales de l'I.H.P*, vol. 5, no. 2, pp. 115–138, 1935.

[GUM 50] GUMBEL E.J., SCHELLING VON H., "The distribution of the number of exceedances", *Annals of Mathematical Statistics*, vol. 21, no. 2, pp. 247–262, 1950.

[GUM 53] GUMBEL E.J., GREENWOOD J.A., DURAND D., "The circular normal distribution, theory and tables", *Journal of the American Statistical Association*, vol. 48, no. 261, pp. 131–152, 1953.

[GUM 54] GUMBEL E.J., "The maxima of the mean largest value and of the range", *Annals of Mathematical Statistics*, vol. 25, no. 1, pp. 76–84, 1954.

[GUM 60] GUMBEL E.J., "Bivariate exponential distributions", *Journal of the American Statistical Association*, vol. 55, no. 292, pp. 698–707, 1960.

[GUM 61] GUMBEL E.J., "Bivariate logistic distributions", *Journal of the American Statistical Association*, vol. 56, no. 294, p.335–349, 1961.

[GUM 67] GUMBEL E.J., MUSTAFI C.K., "Some analytical properties of Bivariate extremal distributions", *Journal of the American Statistical Association*, vol. 62, no. 318, pp. 569–588, 1967.

[GUV 13] GUVENSEN G.M., YILMAZ A.O., "A general framework for optimum iterative blockwise equalization of single carrier MIMO systems and asymptotic performance analysis", *IEEE Transactions on Communications*, vol. 61, no. 2, pp. 609–619, 2013.

[HAD 11] HADZAGIC M., MICHALSKA H., "A Bayesian inference approach for batch trajectory estimation", *IEEE Proceedings of the 14th International Conference on Information Fusion (FUSION)*, pp. 1– 8, 2011.

[HAG 70] HAGER R.O. *Synthetic Aperture Radar Systems Theory and Design*, Academic Press, New York, 1970.

[HAI 96] HAIMOVICH A., "The Eigen canceler, Adaptive Radar by Eigenanalysis Methods", *IEEE Transactions on Aerospace and Electronic systems*, vol. 32, no. 2, pp. 532–542, 1996.

[HAL 00] HALVORSON H, CLIFTON R., "Generic Bell correlation between arbitrary local algebras in quantum field theory", *Journal of Mathematical Physics*, vol. 41, pp. 1711–1717, 2000.

[HAR 12] HARDY L., "The operator tensor formulation of quantum theory", *Phil. Trans. R. Soc. A*, vol. 370, pp. 3385–3417, 2012.

[HAT 02] HATCHER A., *Algebraic Topology*, Cambridge University Press, 2002.

[HEA 13] HEATH R.W., KOUNTOURIS M., BAI T., "Modeling heterogeneous network interference using Poisson Point Processes", *IEEE Transactions on Signal Processing*, vol. 61, no. 16, pp. 4114–4126, 2013.

[HEL 97] HELSINGIUS M., KUOSMANEN P., ASTOLA J., "Nonlinear filter using quantum dot cells". *Electronics Lett.*, vol. 33, no. 20, pp. 1735–1736, 1997.

[HES 67] HESTENES D., "Real spinors fields", *J. Math. Phys.*, vol. 8, no. 4, p.798, 1967.

[HES 84] HESTENES D., SOBCZYK G., *Clifford Algebra to Geometric Calculus, a Unified Language for Mathematics and Physics*, D. Reidel Publishing Company, 1984.

[HIR 13] HIRVENSALO M., "Quantum computing", in RUNEHOV A.L.C., OVLEDO L. (eds), *Encyclopedia of Sciences and Religions*, Springer, 2013.

[HIT 13] HITZER E., NITTA T., KUROE Y., "Applications of Clifford's geometric algebra", *Advances in Applied Clifford Algebras*, vol. 23, no. 2, pp. 377–404, 2013.

[HIT 14] HITZER E., "New developments in Clifford Fourier transforms", *Advances in Applied and Pure Mathematics*, pp. 19–25, 2014.

[HOL 11] HOLEVO A., "Probabilistic and Statistical aspects of quantum theory", *Edizioni della normale*, 2011.

[HOL 87] HOLMES R.B., "Mathematical foundation of signal processing, the role of group theory", Technical Report 781, Lincoln Laboratory MIT, pp. 1–101, 1987.

[HUN 10] HUNG AND CHENG Y., "Cmos nonlinear signal processing circuits", in CHU P.K. (ed.), *Advanced in Solid Sate Circuits Technologies*, Intech, 2010.

[HUR 97] HURT N.E., "Sums and Coding Theory: A review", *Acta Applicandae Mathematicae*, vol. 46, pp. 49–91,1997.

[IAF 96] IAFRATE G.J., STROSCIO M.A., "Application of Quantum based devices: Trends and Challenges", *IEEE Transactions on Electron Devices*, vol. 43, no. 10, pp. 1621–1625, 1996.

[IGL 15] IGLESIAS R., AGUASCI A., FABREGAS X. *et al.*, "Ground based polarimetric Sar interferometry for the monitoring of terrain displacement phenomena. Part II. Applications", *IEEE Journal of Selected Topics in Applied Earth Observations and Remote Sensing*, vol. 8, no. 3, pp. 994–1007, 2015.

[IYE 06] IYENGAR J.R., AMER P.D., STEWART R., "Concurrent multipath transfer using SCTP multihoming over independent end to end paths", *IEEE/ACM Transactions on Networking*, vol. 14, no. 5, pp. 951–964, 2006.

[JAN 96] JANG L., CHAO J., "An information fusion algorithm for data association in multi target tracking", *Data Fusion Symposium*, Adelaide, SA, Australia, 1996.

[JON 08] JONES J.A., "NMR implementations of Gauss sums", *Physics Letters A,* vol. 372, pp. 5758–5799, 2008.

[JON 87] JONES G.A., SINGERMAN D., *Complex Functions, an Algebraic and Geometric Viewpoint*, Cambridge University Press, 1987.

[JOS 05] JOST J., *Riemannian Geometry and Geometric Analysis*, Springer, 2005.

[JUN 13] JUNG S.Y., LEE H.K., KIM S.L., "Worst case user analysis in Poisson Voronoi cells", *IEEE Communicatons Letters*, vol. 17, no. 8, pp. 1580–1583, 2013.

[KAL 02] KALYUZHNY A., KOVTONYUK A., "Basis of the Principal Informative Components for signal estimation problems", *IEEE International Conference on Acoustics, Speech, and Signal Processing (ICASSP),* vol. 2, pp. II1657–II1660, 2002.

[KAL 60] KALMAN R.E., "A new approach to linear filtering and predictive problem", *J. Basic Eng.*, vol. 82, no. 1, pp. 35–45, 1960.

[KAN 09] KANOUN O., "Future prospects for smart sensor systems", *6th International MultiConference on Systems, Signals and Devices*, 2009.

[KAT 06] KATAYAMA T., KELVEY MC. T., SANO A. *et al.*, "Trends in systems and signals", *Committee on Systems and Signal; Annual reviews in Control*, vol. 30, pp. 5–17, 2006.

[KAT 88] KATZ N.M., *Gauss Sums, Klosterman, and Monodromy Groups*, Princeton University Press 1988.

[KAZ 03] KAZHDAN D., POLISHCHIK A., "Fourier transform over finite field and identities between Gauss sums", *Selecta mathematica*, vol. 9, pp. 63–100, 2003.

[KEI 06] KEIMEL K., "Topological cones, foundations for a domain theoretical semantics combining probability and non determinism", *Electronic Notes in Theoritical Computer Science*, vol. 155, pp. 423–443, 2006.

[KEN 11] KENNEDY J., "Particle swarm optimization", *Encyclopedia of Machine Learning*, pp. 760–766, Springer, 2011.

[KIJ 07] KIJEWSKI–CORREA T., KAREEM A., "Nonlinear signal analysis; time frequency perpectives", *Journal of Engineering Mechanics*, vol. 33, no. 2, p. 238, 2007.

[KLE 98] KLEMM R., "Space time adaptive processing principles and applications", *The Institution of Electrical Engineers*, London, 1998.

[KNI 02] KNILL E., LAFLAMME R., BARNUM H. *et al.*, "Introduction to quantum information processing", *ArXiv*, 2002.

[KOL 08] KOLDA T.G., BADER B.W., "Tensor decompositions and applications", *SIAM review*, pp. 1–47, 2008.

[KNU 89] KNUTH D.E., "The art of computer programming", *Addison Wesley Reading*, vol. 1–3, Mass., 1989.

[KOR 92] KOROBOW N.M., "Exponential Sums and their application", SHAKHOV Y., *Mathematics and Its applications (Soviet Series)*, Kluwer Academic Publishers, Dodrecht, p. 80, 1992.

[KOU 02] KOUTSOGIANNIS G.S., SORAGHAN J.J., "Kernel principal component analysis (kpca) for the de-noising of communication signals", *11th Conference European Signal Processing*, vol. 2, pp. 1167–1680, 2002.

[KOU 09] KOUEMOU G., "Exploitation of track accuracy information in fusion technologies for radar target classification using dempster–shafer rules", *12th International Conference on Information Fusion Seattle*, WA, USA, July 6-9 2009.

[KUM 04] KUMAR A., JACOB L., "SCTP vs TCP : Performance Comparison in MANETs", *Proceedings of the 29th Annual IEEE International Conference on Local Computer Networks (LCN)*, 2004.

[KUR 94] KURTHS J., SCHWARZ U., "On nonlinear signal processing", *Proceedings of the Cluster Workshop on Data Analysis Tools*, p. 15, Braunschweig, Germany, 28-30 Sept 1994.

[KVE 03] KVEDALEN E., Signal Processing using the Teager energy operator and other nonlinear operators, PhD Thesis, Oslo University, 2003.

[LAM 13] LAMARE DE R.C., "Adaptive and interative multi branch MMSE decision feedback detection algorithms for multi antenna systems", *IEEE Transactions on Wireless Communications*, vol. 12, no. 10, pp. 5294–5308, 2013.

[LAP 14] LAPLACE P.S., *Théorie analytique des probabilités*, 2nd ed, Paris, Courcier, 1814.

[LAS 93] LASSENBY A., DORAN C, GULL S., "Grassmann calculus, pseudo classical mechanics and geometry algebra", *J. Math. Phys.*, vol. 34, no. 8, pp. 3683–3712, 1993.

[LAT 00b] LATHAUER DE L., MOOR DE B., VANDEWALLE J., "On the best rank-1 and rank-(R_1,R_2 .,…., R_N) approximation of high order tensors", *SIAM J. Matrix Anal. Appl.*, vol. 21, no. 4, pp. 1324–1342, 2000.

[LAT 07] LATHAUWER DE L., "Tensor based techniques for the blind separation of DS–CDMA signals", *Signal Processing*, vol. 87, no. 2, pp. 322–336, 2007.

[LAT 97] LATHAUWER DE L., MOOR DE B., "From matrix to tensor, Multilinear Algebra and Signal Processing", *Mathematics in Signal Processing IV*, Oxford University Press, pp. 1–19, 1997.

[LEE 80] LEE D.T., SCHAECHTER B.J., "Two algorithms for constructing a Delaunay triangulation", *International Journal of Computer & Information Sciences*, vol. 9, no. 3, pp. 219–242, 1980.

[LEF 01] LEFEVRE E., Fusion adaptée d'informations conflictuelles dans le cadre de la théorie de l'évidence, application au diagnostic médical, PhD Thesis, Institut National des Sciences Appliquées de Rennes, Rennes, France, 2001.

[LEN 97] LENT C.S., TOUGAW P.D., "A device architecture for computing with quantum dots", *Proceedings of the IEEE*, vol. 85, no. 4, pp. 541–557, 1997.

[LIN 00] LINATTI J., "On the threshold setting principles in code acquisition of DS-SS signals", *IEEE Journal on Selected Areas in Communications*, vol. 18, no. 1, p.62–72, 2000.

[LI 14] LI R., TURAGA P., SRIVASTAVA A. *et al.*, "Differential geometric representations and algorithms for some pattern recognition and computer vision problems", *Pattern Recognition Letters*, vol. 43, pp. 3–16, 2014.

[LI 15] LI Y., TAO X., LU J., "Effectively lossless subspace appearance model compression using prior information", *Tsinghua Science and Technology*, vol. 20, no. 4, pp. 409–416, 2015.

[LI 90] LI F.K., GOLDESTEIN R.M., "Studies of multibaseline spaceborne interferometric synthetic aperture radar", *IEEE Trans. Geosci. Remote Sensing*, vol. 28, pp. 88–97, 1990.

[LIN 06] LIN Y., VOLKER H., "Application of undersampled Hilbert transform to complex modulated signals", *Instrumentation and Measurement Technology Conference*, pp. 879–882, 2006.

[LUR 04] LURYI S., XU J.M., ZASLAVSKY A., *Future Trends in Microelectronics; the Nano, the Giga, and the Ultra*, Wiley, New York, 2004.

[LOU 99] LOUNEST P., "Clifford algebras and Hestened Spinors", *Foundations of Physics*, vol. 23, no. 9, pp. 1203–1237, 1999.

[MAD 99] MADSEN S.N., "Synthetic aperture radar interferometry; principles and applications", *Manual of Remote Sensing*, vol. 3, chapter. 6, Boston, 1999.

[MAH 07] MAHESH T.S., RAJENDRAN N., PENG X. *et al.*, "Factorizing numbers with gauss sum technique; NMR implementations", *Physical Review A*, vol. 75, pp. 062303–1,062303, 2007.

[MAL 98] MALLET S., *A Wavelet Tour of Signal Processing*, Academic Press, 1998.

[MAN 12] MANGION A.Z., SANGUINETTI G., KADIRKAMANATHAN V., "Variational estimation in spatiotemporal systems from continuous and point process observations", *IEEE Transactions on Signal Processing*, vol. 60, no. 7, pp. 3449–3459, 2012.

[MAR 08] MARSONO M.N., EL-KHARASHI M.W., GEBALI F., "Binary LNS-based naïve Bayes inference engine for spam control: noise analysis and FPGA implementation", *IET Comput. Digit. Tech*, vol. 2, no. 1, p. 56–62, 2008.

[MAR 12] MARVASTI F., AMINI A., HADDADI F. *et al.*, "A unified approach to sparse signal processing", *Eurasip Journal of advances in Signal Processing*, vol. 44, pp. 1–45, 2012.

[MAT 08] MATSUTANI S., "Gauss optics and Gauss sum on an optical phenomena", *Found Phys.*, vol. 38, pp. 758–777, 2008.

[MEN 13] MENNESSON J., SAINT JEAN C.,MASCARILLA L., "Un nouvel ensemble de descripteurs de Fourier Clifford pour les images couleurs", *Les GCFD3*, Hall, 2013.

[MET 54] METER D.V., "Modern statistical approaches to reception in communications theory", *Transactions of the IRE group on Information theory*, vol. 4, no. 4, pp. 119–145, 1954.

[MEY 06] MEYER Y., "Les Ondelettes en 1989", *Lectures Notes in Mathematics*, vol. 1430, pp. 14–25, Springer, 2006.

[MEY 07] MEYER-BASE U., *Digital Signal Processing with Programmable Gate Arrays*, Springer, 2007.

[MIH 10] MIHAILESCU P., VULETESCU V., "Elliptic Gauss Sums and applications to point counting", *Journal of Symbolic Computation*, vol. 45, pp. 825–836, 2010.

[MIS 57] VON MISES R., *Probability, Statistics and Truth*, Dover Publications, 1957.

[MOD 09] MODARRES A.M.A., KAKHKI M.M., "A nonlinear signal processing method for diversity combining using Hammerstein type filter", *International Journal of Computer Science and Network Security*, vol. 9, no. 1, pp. 399–407, 2009.

[MOO 00] MOON T.K., STIRLING W.C., *Mathematical Methods and Algorithms for Signal Processing*, Prentice Hall, 2000.

[MOO 99] MOON T.K., *Mathematical Methods and Algorithms for Signal Processing*, Prentice Hall, 1999.

[MUL 99] MULLER J.M., "Arithmétique des Ordinateurs, ARENAIRE 99", Rapport d'activité INRIA, France, 1999.

[MUT 07] MUTI D., BOURENNANE S., "Survey on tensor signal algebraic filtering", *Signal Processing*, vol. 87, no. 2, pp. 237–249, 2007.

[NAR 82] NARENDRA A., "Dot pattern processing using neighborhoods", *IEEE Transactions on Pattern analysis and machine Intelligence*, vol. PAMI-4, no. 3, pp. 336–343, 1982.

[NG 10] NG H.T., NORI F., "Quantum phase measurement and Gauss sum factorization of large integers in a superconducting circuit", *Physical Review A,* vol. 82, pp. 042317–1,042317, 2010.

[NIE 10] NIELSEN M.A., CHUANG I.L., *Quantum Computation and Quantum Information*, Cambridge University Press, 2010.

[NIM 95] NIMIER V., APPRIOU A., "Utilisation de la Théorie de Dempster-Shafer pour la Fusion d'information", *15ème colloque*, Gretsi Juan les Pins, 18–21 September 1995.

[NIO 10] NION D., SIDIROUPOULOS N.D., "Tensor algebra and multidimensional harmonic retrieval in signal processing for MIMO radar", *IEEE Transactions on Signal Processing*, vol. 58, no. 11, pp. 5693–5705, 2010.

[NYK 02] NYKVIST J., CARR-MOTYCKOVÁ L., "Simulating convergence properties of bgp", *Proc. Computer Communications and Networks*, pp. 124–129, 2002.

[NYQ 24] NYQUIST H., "Certain factors affecting telegraph speed", *The Bell System Technical Journal*, vol. 3, no. 2, pp. 324–346, 1924.

[OBA 13] OBADA A.S.F., HESSIAN H.A., MOHAMMED A.B.A. *et al.*, "A proposal for the realization of universal quantum gates via superconducting qubits inside a cavity", *Annals of physics*, vol. 334, pp. 47–57, 2013.

[OPP 99] OPPENHEIM A., SCHAFER R., BUCK J., *Discrete Time Signal Processing*, Prentice Hall 1999.

[PAD 97] PADALA S.K., PRABHU K.M.M., "Systolic arrays for the discrete Hilbert transform", *IEEE Proceedings Circuits, Devices and Systems*, vol. 144, no. 5, pp. 259–264, 1997.

[PAT 99] PATERSON K.G., "Applications of Exponential Sums in Communications theory", available at: www.hpl.hp.com/techreports/1999/HPL–1999–101.pdf, Hewlett-Packard Company, 1999.

[PEA 01] PEARSON K., "On lines planes of closet fit to systems of points", *Philosophical Magazine series 6*, vol. 2, no. 11, pp. 559–572, 1901.

[PEA 94] PEARSON K., "Contribution of the mathematical theory of evolution", *Philosophical Transactions of the Royal society of London*, vol. 185, pp. 71-110, 1894.

[PEI 00] PEI S.C., DING J.J., "Simplified fractional Fourier transform", *J. Opt. Soc. Am. A*, vol. 17, no. 12, pp. 2355–2367, 2000.

[PEI 98] PEI S.C., MIN Y.M.H., "Discrete fractional Hilbert transform", *IEEE Transactions on Circuits and Systems, II-Analog and Digital Signal Processing*, vol. 47, no. 11, pp. 1307–1311, 1998.

[PEL 14] PELLERIN J., LEVY B., CAUMON G. *et al.*, "Automatic surface remeshing of 3D structural models at specified resolution: a method based on Voronoi diagrams", *Computer & Geosciences*, vol. 62, pp. 103–116, 2014.

[PER 12] PEREYRA M., BATATIA H., "Modeling Ultrasound Echoes in Skin Tissues Using Symmetric α-Stable Processes", *IEEE Transactions on Ultrasonics, Ferroelectrics, and Frequency Control*, vol. 59, no. 1, pp. 60–72, 2012.

[PER 13] PEREZ-CRUZ F., VAERENBERGH S.V., MURILLO-FUENTES J.J. *et al.*, "Gaussian processes for nonlinear signal processing", *IEEE Signal Processing Magazine*, vol. 30, no. 4, p.1, 2013.

[PER 84] PERKINS L.C., SMITH H.B., MOODNEY D.H., "The development of airborne pulse Doppler radar", *IEEE Transactions on Aerospace and Electronic Systems*, vol. 20, no. 3, pp. 292–303, 1984.

[PIE 14] PIEROBON M., AKYILDIZ I.F., "A statistical physical model of interference in diffusion based molecular nanonetworks", *IEEE Transactions on Communications*, vol. 62, no. 6, pp. 2085–2095, 2014.

[PHI 08] PHILIPSON L.L., "The Bayesian anomaly and its practical mitigation", *IEEE Transactions on Reliability*, vol. 57, no. 1, pp. 171–173, 2008.

[PLA 05] PLANAT M., ROSU H., "Mutually unbiased phase states, phase uncertainties, and Gauss sums", *Eur. Phys.J.D.*, vol. 36, pp. 133–139, 2005.

[POL 84] POLYDOROS A., "A unified approach to serial search spread spectrum code acquisition-Part I: General theory", *IEEE Transactions on Communications*, vol. 32, pp. 542–549, 1984.

[POT 83] POTTER J.M., ANDERSON B.D.O., "Statistical inference with partial prior information", *IEEE Transactions on Information Theory*, vol. IT-29, no 5, pp. 688–695, 1983.

[PRA 08] PRANDONI P, VETTERLI M., *Signal Processing for Communications*, CRC Press, 2008.

[PRO 95] PROAKIS J., *Digital Communications*, McGraw-Hill, 1995.

[PSA 99] PSAROMILIGKOS I.N., BATALAMA S.N., PADOS D.A., "On adaptive minimum probability of error linear filter receivers for DS-CDMA channels", *IEEE Transactions on Communications*, vol. 47, no. 7, pp. 1092–1102, 1999.

[QU 07] QU H.B., HU B.G., "Variational Bayes inference for generalized associative functional networks", *Proceedings of International Joint Conference on Neural Networks*, Orlando, USA, pp. 184–189, 2007.

[RAE 99] RAE H., "Performance analysis of a double-dwell serial search technique for cellular CDMA networks in the case of multiple pilot signals", *IEEE Transactions on Vehicular Technology,* vol. 48, no. 6, November 1999.

[RAH 11] RAHMAN M.Z.U., REDDY D.V.R.K., SANGEETHA Y, "Stationary and non stationary noise removal from cardiac signals using a constrained stability least mean square algorithm", *International Conference on Communications and Signal Processing,* pp. 485–488, 2011.

[RAR 41] RARITA W., SCHWINGER J., "On a theory of particles with half integral spin", *Physical Review*, vol. 60, p.61, 1941.

[RIC 01] RICHARDSON P.G., "STAP covariance matrix structure and its impact on clutter plus jamming suppression solutions", *Electronics Letters*, vol. 37, no. 2, pp. 118–119, 2001.

[ROC 04] ROCKMORE D.N., HEALY JR D.M., *Modern Signal Processing,* Cambridge University Press, 2004.

[ROT 70] ROTTERDAM A.V., "Limitations and difficulties in signal processing by means of principal component analysis", *IEEE transactions on Biomedical Engineering,* p. 268, 1970.

[RUI 07] RUIZ-ALZOLA J., WESTIN C., "Tensor signal processing", *Signal Processing,* vol. 87, no. 2, pp. 217–219 ,2007.

[SAB 85] SABURI Y., "Fundamental properties of modified Fourier Hyperfunctions", *Tokyo J Math.*, vol. 8, no. 1, pp. 231–273, 1985.

[SAT 59] SATO M., "Theory of Hyperfunctions I", *Journal of the Faculty of Science, Sect1, mathematics, astronomy, physics, chemistry,* vol. 8, no. 1, pp. 139–193, 1959.

[SAT 60] SATO M., "Theory of Hyperfunctions II", *Journal of the Faculty of Science, Sect1, mathematics, astronomy, physics, chemistry,* vol. 8, no. 2, pp. 387–437, 1960.

[SHA 13] SHAPIRA P., "Hyperbolic systems and propagation on causal manifolds", *Letters in Mathematical Physics*, vol. 103, pp. 1149–1164, 2013.

[SCH 51] SCHWARTZ L., *Théorie des Distributions*, 1st ed., Hermann, 1951.

[SHA 48] SHANNON C.E., "A mathematical theory of communication", *Bell Syst. Tech. J.*, vol. 27, no 1948, pp. 379–457 & 623–656.

[SHA 49] SHANNON C.E., "Communication in the presence of noise", *Proc. IRE,* vol. 37, pp. 10–21, 1949.

[SHA 75] SHAMAS M.I., HOEY D., "Closest point problems", *IEEE 16th Annual symposium on Foundations of Computer Sciences*, pp. 151–162, 1975.

[SHA 76] SHAFER G., *A Mathematical Theory of Evidence*, Princeton University Press, New Jersey, 1976.

[SHO 10] SHOJAE F., DEHESTANI M., "The simulation of virus life cycle quantum gates", *Computer in Biology and Medicine,* vol. 40, pp. 359–362, 2010.

[SIC 15] SICA F., REALE D., POGGI G. *et al.*, "Non local adaptive multilooking in SAR multipass differential interferometry", *IEEE Journal of Selected topics in applied Earth Observations and Remote Sensing*, vol. 8, no. 4, pp. 1727–1742, 2015.

[SID 05] SIDKY E.Y., PAN X., "Recovering a compactly supported function from knowledge of its Hilbert transform on a finite interval", *IEEE Signal Processing Letters*, vol. 12, no. 2, pp. 97–100, 2005.

[SLE 74] SLEPIAN D., WOLF J.K., "Noiseless coding of correlated information sources", *IEEE Transactions on Information theory*, vol. IT-19, no. 4, pp. 471–480, 1973.

[SME 90] SMETS P., HENRION M., KANAL L.N.," Constructing the pignistic probability function in a context of uncertainty", *Uncertainty in Artificial Intelligence*, Amsterdam, North-Holland, vol. 5, pp. 28–40. 1990.

[SOM 15] SOMMEN F., SCHEPPER DE H., *Introductory Clifford analysis. Operator Theory*, Springer, Basel, 2015.

[SOS 06] SOSULIN Y.G., KOSTROV V.V., "Estimation-correlation-compensation signal processing in the presence of interferences, a review", *Journal of Communications technology and electronics*, vol. 51, no. 9, pp. 967–1004, 2006.

[STO 05] STOICA P., MOSES R., *Spectral Analysis of Signals*, Prentice Hall, 2005.

[SU 10] SU Y.T., WONG K.T., "Linear MMSE Estimation of Large-Magnitude Symmetric Levy-Process Phase-Noise", *IEEE Transactions on Communications*, vol. 58, no. 12, pp. 3369–3374, 2010.

[SU 11] SU Y.T., WONG K.T., "Algorithmic extensions of Su-Wong-Ho linear MMSE estimator for large-magnitude Levy-process phase-noise", *Electronics Letters*, vol. 47, no. 11, p. 653-654, 2011.

[SZU 13] SZUMNIAK P., Design and computer simulations of nanodevices with applications to quantum computing, PhD Thesis, University of Cracovie, Poland, 2013.

[TAK 05] TAKEUCHI J.I., AMARI S.I., "α parallel prior and its properties", *IEEE Transactions on Information Theory*, vol. 51, no. 3, pp. 1011–1023, 2005.

[TAK 06] TAKAHASHI H., "Bayesian approach to a definition of random sequences and its applications to statistical inference", *ISIT*, Seattle, USA, pp. 2180–2184, 2006.

[TAY 12] TAYLOR D.B., DHILLON H.S., NOVLAN D. *et al.*, "Pairwise intercation processes for modeling cellular network topology", *IEEE Global Communications Conference*, pp. 4524–4529, 2012.

[THA 13] THAKOR N.V., SHERMAN D.L., "EEG signal processing, theory and applications", *Neural Engineering*, Springer, 2013.

[TOD 14] TODD A., BIHAN LE N., SANGWINE S.J., *Quaternion Fourier Transform for Signal and Image Processing*, ISTE, London and John Wiley & Sons, New York, 2014.

[ULA 04] ULAM S.M., *Problems in Modern Mathematics,* Dover Publications, 2004.

[VAL 14] VALENTE R.A., MARINELLO J.C., ABRAO T., "LR aided MIMO detectors under correlated and imperfectly estimated channels", *Wireless Pers. Commun*, vol. 77, pp. 173–196, 2014.

[VAN 05] VANKAMAMIDI V., OTTAVI M., LOMBARDI F., "A line based parallel memory for QCA implementation" *IEEE Transactions on Nanotechnology*, vol. 4, no. 6, pp. 690–698, 2005.

[VAR 89] VARANASI M.K., "Parametric generalized Gaussian density estimation", *Journal of the Acoustical Society of America*, vol. 86, no. 4, pp. 1404–1415, 1989.

[VIN 08] VINCE J., *Geometric Algebra for Computer Graphics*, Springer, 2008.

[VER 14] VERVLIET N., DEBALS O., SOBERT L. *et al.*, "Breaking the curse of dimensionality using decompositions of incomplete tensors", *IEEE Signal Processing Magazine*, pp. 71–79, 2014.

[VIT 67] VITERBI A.J., "Error bounds for convolutional codes and asymptotically optimum decoding algorithm", *IEEE Transactions on Information Theory*, pp. 260–269, 1967.

[VIT 71] VITERBI A.J., "Convolutional codes and their performance in communication systems", *IEEE Trans. Commun.*, vol. COM-19, pp. 751–772, October 1971.

[VIT 95] VITERBI A.J., *CDMA: Principles of Spread Spectrum Communications*, Addison-Wesley, 1995.

[WAN 09] WANG Z., PEI X., "Application of Principal Component Extraction technique in processing cloud images from Chinese FY-1 satellite", *IASP International Conference on Image Analysis and Signal Processing*, pp. 18–21, 2009.

[WAL 73] WALD A., *Sequential Analysis*, Dover Publications,1973.

[WAN 11A] WANG J., XIA Y., FENG D.D., "Differential evolution based variational bayes inference for brain PET-CT image segmentation", *International Conference on Digital Image Computing Techniques and applications*, pp. 330–334, 2011.

[WAN 11b] WANG Q., KHANG F., "Bayes inference for loss and risk function of exponential distribution based on progressively type II censoring", *International Conference on Multimedia technology*, pp. 2008–2011, 2011.

[WAT 01] WATARI M., ERNST T., WAKIKAWA R. *et al.*, "Routing optimization for nested mobile networks, special section on mobile multimedia communications", *IEICE Trans. Commun.*, vol. E89–B, no. 10, 2006.

[WEI 71] WEINSTEIN S., EBERT P., "Data transmission by frequency division multiplexing using the discrete Fourier Transform", *IEEE Transactions on Communications Technology*, vol. 19, no. 5, pp. 628–634, 1971.

[WES 94] WESTIN C.F., A tensor framework for multidimensional signal processing, PhD Thesis, Linköping University, 1994.

[WIC 12] WICKS M.C., VELA R., MONTE L.L., "Range Doppler angle ambiguity function analysis in modern radar", *IEEE International symposium on antennas and Propagation*, p.1, 2012.

[WIE 49] WIENER N., *Extrapolation and Smoothing of Stationary Time Series with Engineering Applications*, MIT Press, 1949.

[WIL 13] WILDEMEERSCH M., QUEK T.Q.S., SLUMP C.H. *et al.*, "Cognitive small cell networks; energy efficiency and trade-offs", *IEEE Transactions on Communications*, vol. 61, no. 9, pp. 4016–4029, 2013.

[WHI 07] WHITNEY M., ISAILOVIC N., PATEL Y. *et al.*, "Automated generation of layout and control of quantum circuit", *Proceedings of the 4th International conference on Computing frontiers*, ACM New York, pp. 83–94, 2007.

[WOO 06] WOOLRICH M.W., BEHRENS T.E., "Variational Bayes inference of a spatial mixture models for segmentation", *IEEE Transactions on Medical Imaging*, vol. 25, no. 10, pp. 1380–1391, 2006.

[XIA 13] XIANG L., GE X., WANG C.X. *et al.*, "Energy efficiency evaluation of cellular networks based on spatial distributions of traffic load and power consumption", *IEEE transactions on Wireless Communications*, vol. 12, no 3, pp. 961–973, 2013.

[XIE 12] XIE S., KRISHNAN S., LAWNICZAK A.T., "Sparse principal component extraction and classification of long term biomedical signals. Computer-Based Medical Systems (CBMS)", *25th International Symposium*, pp. 1–6, 2012.

[XU 09] XU J.,HU S., SHEN Z., "Fault detection for process monitoring using improved kernel principal component analysis", *International conference on Artificial intelligence and Computational Intelligence*, pp. 334–338, 2009.

[ZAN 04] ZANARDI P., LIDAR D.A., LLOYD S., "Quantum tensor product structures are observable induced", *Phys. Rev. Lett*, vol. 92, pp. 6–13, 2004.

[ZAY 98] ZAYED A.I., "Hilbert transform associated with the fractional Fourier transform", *IEEE Signal processing Letters*, vol. 5, no. 8, pp. 206–209, 1998.

[ZEB 86] ZEBKER H.A., GOLDESTEIN R.M., "Topographic mapping from interferometric SAR observations", *J. Geophys. Res.*, vol. 91, pp. 4993–4999, 1986.

[ZHA 13] ZHAO Y., FENG J., ZHANG B.C. *et al.*, "Current progress in sparse signal processing applied to radar imaging", *Science China Technological Sciences*, vol. 56, no. 12, pp. 3049–3054, 2013.

[ZHU 07] ZHU X., ZHANG L, WEI J. *et al.*, "Application of wavelets and principal component analysis to process quantitative feature extraction", *IEEE International Conference on Control and Automation*, China, pp. 2609–2614, 2007.

[ZOU 13] ZOU J., KUSYK J., UMIT UYAR M. *et al.*, "Bio inspired and Voronoi based algorithms for self positioning autonomous mobiles nodes", *IEEE Military Communications Conference*, pp. 1–6, 2013.

Index

A, B

C

D

P, Q

R, S

T, U

V, W

Other titles from

iSTE

in

Networks and Telecommunications

2016

CAMARA Daniel, NIKAEIN Navid (Eds.)
Wireless Public Safety Networks 2: A Systemic Approach

CHIASSERINI Carla Fabiana, GRIBAUDO Marco, MANINI Daniele
Analytical Modeling of Wireless Communication Systems
(Stochastic Models in Computer Science and Telecommunication Networks Set – Volume 1)

PEREZ André
VoLTE and ViLTE

RAMOS Angel, LAZARO Antonio, GIRBAU David, VILLARINO Ramon
RFID and Wireless Sensors using Ultra-Wideband Technology

SENOUCI Mustapha Reda, MELLOUK Abdelhamid
Deploying Wireless Sensor Networks

2015

BENSLAMA Malek, KIAMOUCHE Wassila, BATATIA Hadj
Connections Management Strategies in Satellite Cellular Networks

BENSLAMA Malek, BATATIA Hadj, BOUCENNA Mohamed Lamine
Ad Hoc Networks Telecommunications and Game Theory

BERTHOU Pascal, BAUDOIN Cédric, GAYRAUD Thierry, GINESTE Matthieu
Satellite and Terrestrial Hybrid Networks

CAMARA Daniel
Bio-inspired Networking

CAMARA Daniel, NIKAEIN Navid (Eds.)
Wireless Public Safety Networks 1: Overview and Challenges

CUADRA-SANCHEZ Antonio, ARACIL Javier
Traffic Anomaly Detection

JARRY Pierre, BENEAT Jacques N.
Digital Communications

LE RUYET Didier, PISCHELLA Mylène
Digital Communications 1: Source and Channel Coding

PEREZ André
LTE and LTE Advanced: 4G Network Radio Interface

PISCHELLA Mylène, LE RUYET Didier
Digital Communications 2: Digital Modulations

PUJOLLE Guy
Software Networks

2014

ANJUM Bushra, PERROS Harry
Bandwidth Allocation for Video under Quality of Service Constraints

BATTU Daniel
New Telecom Networks: Enterprises and Security

BEN MAHMOUD Mohamed Slim, GUERBER Christophe, LARRIEU Nicolas, PIROVANO Alain, RADZIK José
Aeronautical Air–Ground Data Link Communications

BITAM Salim, MELLOUK Abdelhamid
Bio-inspired Routing Protocols for Vehicular Ad-Hoc Networks

CAMPISTA Miguel Elias Mitre, RUBINSTEIN Marcelo Gonçalves
Advanced Routing Protocols for Wireless Networks

CHETTO Maryline
Real-time Systems Scheduling 1: Fundamentals
Real-time Systems Scheduling 2: Focuses

EXPOSITO Ernesto, DIOP Codé
Smart SOA Platforms in Cloud Computing Architectures

MELLOUK Abdelhamid, CUADRA-SANCHEZ Antonio
Quality of Experience Engineering for Customer Added Value Services

OTEAFY Sharief M.A., HASSANEIN Hossam S.
Dynamic Wireless Sensor Networks

PEREZ André
Network Security

PERRET Etienne
Radio Frequency Identification and Sensors: From RFID to Chipless RFID

REMY Jean-Gabriel, LETAMENDIA Charlotte
LTE Standards
LTE Services

TANWIR Savera, PERROS Harry
VBR Video Traffic Models

VAN METER Rodney
Quantum Networking

XIONG Kaiqi
Resource Optimization and Security for Cloud Services

2013

ASSING Dominique, CALÉ Stéphane
Mobile Access Safety: Beyond BYOD

BEN MAHMOUD Mohamed Slim, LARRIEU Nicolas, PIROVANO Alain
Risk Propagation Assessment for Network Security: Application to Airport Communication Network Design

BEYLOT André-Luc, LABIOD Houda
Vehicular Networks: Models and Algorithms

BRITO Gabriel M., VELLOSO Pedro Braconnot, MORAES Igor M.
Information-Centric Networks: A New Paradigm for the Internet

BERTIN Emmanuel, CRESPI Noël
Architecture and Governance for Communication Services

DEUFF Dominique, COSQUER Mathilde
User-Centered Agile Method

DUARTE Otto Carlos, PUJOLLE Guy
Virtual Networks: Pluralistic Approach for the Next Generation of Internet

FOWLER Scott A., MELLOUK Abdelhamid, YAMADA Naomi
LTE-Advanced DRX Mechanism for Power Saving

JOBERT Sébastien *et al.*
Synchronous Ethernet and IEEE 1588 in Telecoms: Next Generation Synchronization Networks

MELLOUK Abdelhamid, HOCEINI Said, TRAN Hai Anh
Quality-of-Experience for Multimedia: Application to Content Delivery Network Architecture

NAIT-SIDI-MOH Ahmed, BAKHOUYA Mohamed, GABER Jaafar, WACK Maxime
Geopositioning and Mobility

PEREZ André
Voice over LTE: EPS and IMS Networks

2012

AL AGHA Khaldoun
Network Coding

BOUCHET Olivier
Wireless Optical Communications

DECREUSEFOND Laurent, MOYAL Pascal
Stochastic Modeling and Analysis of Telecoms Networks

DUFOUR Jean-Yves
Intelligent Video Surveillance Systems

EXPOSITO Ernesto
Advanced Transport Protocols: Designing the Next Generation

JUMIRA Oswald, ZEADALLY Sherali
Energy Efficiency in Wireless Networks

KRIEF Francine
Green Networking

PEREZ André
Mobile Networks Architecture

2011

BONALD Thomas, FEUILLET Mathieu
Network Performance Analysis

CARBOU Romain, DIAZ Michel, EXPOSITO Ernesto, ROMAN Rodrigo
Digital Home Networking

CHABANNE Hervé, URIEN Pascal, SUSINI Jean-Ferdinand
RFID and the Internet of Things

GARDUNO David, DIAZ Michel
Communicating Systems with UML 2: Modeling and Analysis of Network Protocols

LAHEURTE Jean-Marc
Compact Antennas for Wireless Communications and Terminals: Theory and Design

RÉMY Jean-Gabriel, LETAMENDIA Charlotte
Home Area Networks and IPTV

PALICOT Jacques
Radio Engineering: From Software Radio to Cognitive Radio

PEREZ André
IP, Ethernet and MPLS Networks: Resource and Fault Management

TOUTAIN Laurent, MINABURO Ana
Local Networks and the Internet: From Protocols to Interconnection

2010

CHAOUCHI Hakima
The Internet of Things

FRIKHA Mounir
Ad Hoc Networks: Routing, QoS and Optimization

KRIEF Francine
Communicating Embedded Systems / Network Applications

2009

CHAOUCHI Hakima, MAKNAVICIUS Maryline
Wireless and Mobile Network Security

VIVIER Emmanuelle
Radio Resources Management in WiMAX

2008

CHADUC Jean-Marc, POGOREL Gérard
The Radio Spectrum

GAÏTI Dominique
Autonomic Networks

LABIOD Houda
Wireless Ad Hoc and Sensor Networks

LECOY Pierre
Fiber-optic Communications

MELLOUK Abdelhamid
End-to-End Quality of Service Engineering in Next Generation Heterogeneous Networks

PAGANI Pascal *et al.*
Ultra-wideband Radio Propagation Channel

2007

BENSLIMANE Abderrahim
Multimedia Multicast on the Internet

PUJOLLE Guy
Management, Control and Evolution of IP Networks

SANCHEZ Javier, THIOUNE Mamadou
UMTS

VIVIER Guillaume
Reconfigurable Mobile Radio Systems

Lightning Source UK Ltd.
Milton Keynes UK
UKHW022224240519
343290UK00003B/54/P

9 781848 219250